# Hear All the Action

by Van Waterford

Howard W. Sams & Co., Inc.
4300 WEST 62ND ST. INDIANAPOLIS, INDIANA 46268 USA

# Preface

During the past couple of years, almost every citizen in this country has become aware of radiocommunications through the tremendous popularity of Citizens Band radio.

However, CB radio, amateur radio, and any other two-way radio communications requires at least one active element on the part of the participants—talking. But there are in this country—and in many other areas of the world—a large number of radio enthusiasts, who do not like to "transmit," but prefer simply to listen to the world of events from far away or nearby.

This group includes people called DXers—armchair listeners. Radio eavesdropping as a hobby has been around for many, many years. In this communications-oriented world, the amateur listener ranks continue to expand. There are those who listen to shortwave radio only to hear those strange sounds from far-off lands, or those who listen only to amateurs talking to each other from one side of the globe to the other. Then, there are the "actionbanders," those who listen to the police and fire department frequencies in their own neighborhood.

As with other hobbies, knowledge is essential to further the advance of the enthusiast. This book is, I hope, such a tool. It is addressed to all who are interested in listening to overseas broadcasts, news events, music (exotic and otherwise), language lessons, history, culture, and the voices of police officers chasing criminal suspects.

Following a brief history of DXing in the Introduction, the reader is presented with basic terms and fundamentals of shortwave radio communications. Chapter 1 deals with such things as frequency, wavelength, propagation, vhf and uhf frequency bands, etc. In Chapter 2 you will find out how to shop for DX receivers and what electronic specifications are important to the DXer. This chapter also covers how to set up your DX station. Chapter 3 looks at the various accessories and operating aids including: headphones, antenna tuners, audio filters, logbooks, and maps. A variety of receiving antennas and how to construct them with step-by-step details is presented in Chapter 4. QSL reports—a fascinating part of DXing—are the subject of Chapter 5 which also tells you where to send for QSL reports. Chapter 6 covers public service band monitoring and contains several special codes used by police and emergency communicators. Selecting your equipment is the topic of Chapter 7 and deals with such equipment as crystal-controlled receivers and computerized scanners. The Appendix contains a glossary, a list of common abbreviations, and miscellaneous codes sometimes heard in radio communications.

The book was started at the California beaches and finished 10,000 feet high in the Rocky Mountains, where sounds from the world were captured by the thousands. The writing was interspersed by days of listening to sheriff's officers pursuing escaped convicts through abandoned mines and shacks—heavy underbrush and steep hills.

Several individuals and organizations contributed to the writing of this book with encouragement, help, and information, especially my wife, who went through days and nights of lonely vigil. Others include the Association of North American Radio Clubs; Rudolf Heim of the European DX Council, who was such a gracious host when we visited him in Bochum, West Germany; all the DX Clubs for the information they sent me; and all the manufacturers of communication receivers, scanning monitors, and accessories for the material and photographs they submitted.

VAN WATERFORD

# Contents

# History of DXing

It was in December 1901 when Guglielmo Marconi made the first transatlantic broadcast. There were others, however, before Marconi who were able to send wireless messages across great distances.

As early as 1866, Mahlon Loomis experimented with electricity and succeeded in sending a message 20 miles through the air. Loomis' system involved the conducting of large currents of electricity through salt-water puddles and kites flown by wires. He connected a galvanometer to one kite and charged another kite with electricity, causing the galvanometer to deflect. Although this was not a true radio signal, it did constitute a man-made signal.

Another experimenter was Nathan Stubblefield of Murray, Kentucky. He put together a large coil that was attached to a telephonelike mouthpiece. By grounding his equipment he was able to conduct actual voice communications without wires. The year was 1888. Stubblefield's patent was never commercialized, however. After Congress had watched a demonstration of his "Vibrating Telephone," it decided to allocate $50,000 to further development of the apparatus. The money so allocated never arrived. Stubblefield, a heartbroken and penniless farmer, died in 1928.

In the meantime, Marconi came to the United States and developed his wireless machine further. He set up a station on Cape Cod in Massachusetts and started operating in January of 1903 by sending wireless code messages to Europe for 50 cents a word.

The first commercial broadcast was transmitted from San Jose, California. The station, first licensed as KQW, started in 1909 and was founded by Dr. Charles David Herrold who distributed crystal sets to friends and relatives in the area. They could then listen to music, news, and other programming from the station. KQW later moved to San Francisco and the call letters were changed to KCBS.

Broadcasting developed by leaps and bounds. And so today, most of the countries in the world can be heard broadcasting on the shortwave bands. Although shortwave radio in America is, for most listeners, a hobby and a source of entertainment and information beyond local and national sources, such as newspapers and tv, it is the *prime* communications medium in Afro, Asian, and South American countries. Shortwave is the key to national unity for many countries. It is used for educational purposes, news of government activities, etc. Shortwave radio is often the only way a government can reach its people—it is their standard domestic broadcast band. The am and fm bands, so widely used in the developed countries, are little used by the developing nations. This means that their domestic broadcasting occurs on the same general range of frequencies as the international broadcasting conducted by the developed nations. The result: crowded shortwave bands, but also enormous opportunities for the DXer to receive more stations than ever before.

In 1979, the International Telecommunications Union will hold a World Administrative Conference. The ITU is expected to decide on the size and structure of the international shortwave broadcast bands for the remainder of this century.

Conference discussions are expected to be difficult and sensitive. Yet, overall, the future for international broadcasting appears better than it has for some time. There are indications that some positive correlation exists between the quality of broadcast signals and the size of the shortwave listening (SWL) audience, leading to the suggestion that international broadcasting in the 1980s will become more attractive to you as a DXer.

SWL may never reach the fantastic heights of CB (citizens band) popularity—and then, too, it may. There is an enormous boom in shortwave listening in Japan which may extend to other countries. One result of this boom is the availability of a wider range of receivers. Not too long ago, the choice—especially of medium-priced receivers—was limited. Now, however, there are a number of receivers manufactured in Japan and the U.S. that are priced in the midrange of $100 to $400. So, all the elements are there for you to become a DXer.

The only thing remaining is to explain to your friends and relatives what a DXer is. Tell them simply that you "just listen to the action around town and the world." It is not really a hostile world that the DXer faces, but just a world in which the average person thinks that a "killuh-cycle" is a souped-up Hell's Angels Harley Davidson. Do not become a DX-hermit. Rather, explain to everyone you meet that DXing is the most interesting, entertaining, educational, and enjoyable hobby around. . . . You can travel the globe without ever leaving home, and you cannot beat the price. . . . That DXing is a terrific way to experience the music, the culture, and the customs of all sorts of far-off places. . . . That you get a thrill from just knowing that the signal you have just tuned in has spanned half the world. . . . That you are keeping abreast of what is happening in your community by monitoring police and fire calls.

Now on to the important concepts about which you will want to have a better understanding, such as radio-wave bands, frequency, wavelength, wave propagation, antennas, accessories, etc. Some knowledge of these things will help make your shortwave listening much more enjoyable. So, let's get to it.

# CHAPTER 1

# The Radio-Wave Bands

Two words that are used throughout this book may at first seem interchangeable. These two words are frequency and wavelength. And they cannot be used interchangeably. In fact, each is exactly the reciprocal of the other, mathematically speaking. The term "shortwave listener," abbreviated SWL, may be a little bit confusing, therefore. In the strictest sense, SWL refers to a person who listens only to the actual shortwave frequencies on a radio receiver. Seemingly, this would not include anyone who listens to the range of frequencies that are found on the common home or car radio. This frequency range generally is known as "the medium waves." In recent years, however, SWL sometimes has been construed to include anyone who listens to any frequency as long as it can be tuned in on a radio receiver. This would include listening to the standard broadcast band and even that relatively unknown range of frequencies called "the long waves." It might be well, before proceeding any further, to have an explanation of the terms frequency and wavelength.

All radio-frequency signals have some things in common. Whether amplitude modulated (am) or frequency modulated (fm), the signals travel at the speed of light—about 300,000,000 meters per second or approximately 186,000 miles per second. (By international agreement, radio wavelengths are measured in meters rather than in yards or feet.)

Radio waves are alternating waves; that is, the strength of the wave pulsates from zero to maximum and back to zero, first in one direction, and then from zero to maximum and back to zero, in the opposite direction. This does not mean that the wave itself changes direction as it moves through space, but only that the strength of the wave changes. In other words, the peaks and dips of a radio signal pass any given point at a speed of 300,000,000 meters per second.

Because the radio wave occupies both time and space, the wavelength, frequency, and the speed of light can all be expressed by a formula that translates wavelength into frequency and vice versa, as will be explained in the following paragraphs.

## FREQUENCY

It was mentioned earlier that radio waves are *alternating* waves consisting of peaks and dips. The number of peaks measured is called the "frequency of the signal." One peak and one dip is called a "cycle." See Fig. 1-1. One cycle per second is one *hertz*—abbreviated Hz. The radio wave shown in Fig. 1-1 goes through 3 cycles in 1 second, i.e., the frequency is 3 Hz.

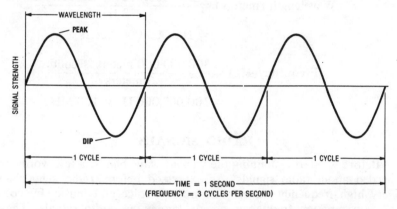

Fig. 1-1. Radio waves are alternating waves.

Radio waves are very high frequency waves. They are measured in thousands and millions of hertz. Rather than to express radio frequencies as 500,000 Hz, 10,000,000 Hz, etc., the prefixes "kilo" equals 1,000 and "mega" equals 1,000,000 are used. Substituting "k" as the abbreviation for kilo, 500,000 Hz is expressed as 500 kHz. In the same manner, mega is abbreviated "M" and 10,000,000 Hz is expressed as 10 MHz.

## WAVELENGTH

Wavelength and frequency are directly related—not interchangeable. Wavelength, as shown in Fig. 1-1, is the distance traveled by the radio wave in the time required for the wave to complete one cycle. In other words, it is the physical length of one cycle. Wavelength is expressed in meters and if the frequency of a wave is known, its wavelength can be determined by dividing the velocity of the wave by the frequency. For example, consider 30 million peaks and dips passing a specified point each second. The distance from one peak to the next peak, therefore, is 10 meters which is equal to the length of one radio wave. Thus, a 10-meter radio wave has a frequency of 30 MHz. Let's take another example. The wave peaks of the 600-kHz broadcast station develop at a rate of 600,000 cycles per second. Frequency is 600 kHz and the velocity is constant at 300 million meters per second. The wavelength (distance between peaks) is 500 meters—velocity of 300,000,000 divided by the frequency of 600,000 equals wavelength of 500.

The mathematical equations for wavelength or frequency are:

$$\text{Wavelength (meters)} = \frac{300,000,000 \text{ meters/second}}{3 \text{ Hz}}$$

$$= 100,000,000 \text{ meters}$$

$$\text{Frequency(Hz)} = \frac{300,000,000 \text{ meters/second}}{3 \text{ meters}}$$

$$= 100,000,000 \text{ Hz} = 100 \text{ MHz}$$

## RADIO SIGNALS

Before we start examining the various wave bands, a few words are needed about radio signals. As expressed before, radio signals are very high frequency waves. The human voice and music, by comparison, are low-frequency signals known as *audio* signals. These audio signals range from about 20 Hz to 15,000 Hz. Because they are of such low frequencies and other technical considerations, they cannot be transmitted over long distances like radio-frequency signals. Therefore, at the transmitting station, the audio signal (voice and music) is superimposed on the radio signal by the electronic process called *modulation*. The combined signal is then radiated into space by the transmitting antenna, picked up by another antenna a few thousandths of a second later, and fed to a receiver. Within the receiver, the audio signal is separated from the radio signal by the electronic process of *demodulation* and then amplified and fed to the speaker or headphones. The strength of the signal received by the antenna is measured in microvolts (millionths of a volt). To make sure that an antenna receives the strongest signal possible, transmitting stations commonly use directional antennas to beam (aim) the signal in the desired direction.

You cannot see the radio signals leaving the antenna, but there are instruments that can measure and describe them. Some signals may be controlled and there are others which are beyond our power to control.

There is no mystery in DXing and nothing magical about the phenomena that bring distant shortwave signals to your receiver. But still, there are some SWLers who persist in the notion that the propagation of radio signals is a deep, dark mystery.

True, there are things we do not know regarding when and why DX signals can be heard. But the basis has long been known. And if you take the time to learn a little bit about propagation, you will be better equipped to tune in to the right frequencies at the right times. It will give you the best chance to hear the DX station you are looking for.

## PROPAGATION

Without propagation, there would be no such thing as DXing. All listening would be limited to stations located not too far from the listener's receiver. Luckily, though, we have the ionosphere which allows propagation.

As the radio signal escapes into space the angle and direction of the signal can be controlled by making changes in the electrical characteristics of the antenna and by changing the height of the antenna above the surface of the ground. After the wave leaves the antenna system, however, we are powerless to control it. Natural phenomena

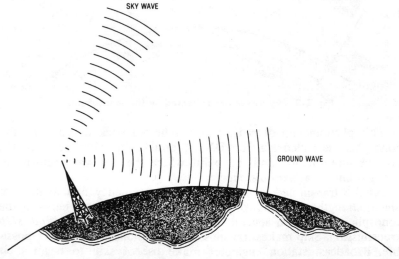

**Fig. 1-2. Radio-wave transmission.**

also have an effect on radio waves. The science concerned with these effects is called *propagation*.

When a radio wave is transmitted, various parts of it travel from the transmitting antenna to different parts of the "air," as shown in Fig. 1-2. One part, called the *ground wave* travels along the surface of the earth. Commercial broadcasting stations make use of this ground wave, since it has proven to be a very reliable means for communicating over medium distances. Ground-wave coverage in the broadcast band may range up to 500 miles, the coverage increases as the frequency diminishes. Large portions of the earth may be covered by the ground wave at frequencies in the 100 to 200 kHz region (low frequency). Ground-wave propagation is also used for television and fm broadcasting.

Another part of the radio wave, called the *sky wave,* travels upward and outward into space. The sky wave makes shortwave listening possible. As a sky wave travels upward into space, it encounters the *ionosphere,* a region consisting of layers of ionized gases about 60 to 200 miles above the earth. (See Fig. 1-3.) The sky wave is reflected back to earth by the ionosphere. This layer is formed as energy from the sun strikes particles of the air and ionizes them, releasing free electrons from the particles, causing a blanket of electrons to be formed.

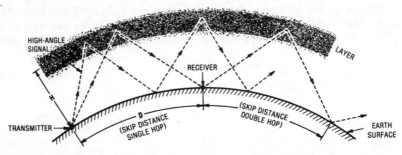

**Fig. 1-3. Sky waves are reflected by the ionosphere.**

This phenomenon of radio waves reflected back to earth by the ionosphere is called *skip* by DXers, and the distance between the transmitting antenna and the point where the sky wave returns to earth is known as *skip distance.*

As the transmitted signal depends on frequency and angle, not signal strength, the signal often bounces back and forth between the ionosphere and earth several times. This is known as *multiple skip* transmission. Skip makes transmission over thousands of miles possible. Broadcast station engineers make use of skip to beam their broadcast to certain target areas. They do this by controlling the

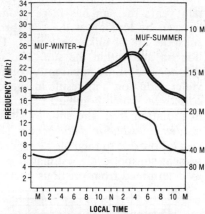

Fig. 1-4. Maximum Usable Frequency
chart for use in the U.S.A.

angle, called the *critical angle,* at which a radio signal of a certain
frequency strikes the ionosphere. They also make use of the *maximum
usable frequency* (MUF): the highest radio frequency that will be
reliably reflected back to earth from the ionosphere. And since the
ionosphere is directly affected by the *sunspot cycle,* so too is the MUF.
The MUF is highest during peak sunspot years. Fig. 1-4 shows a
chart for Maximum Usable Frequency for use in the United States.
Note the difference between MUF-winter and MUF-summer. A
30-MHz frequency at noon local time is best to utilize skip conditions
in the winter, for example.

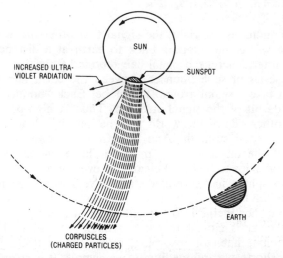

Fig. 1-5. Increase in radiation from the sun is responsible for sudden
ionospheric disturbances.

The sunspot intensity varies over an eleven-year cycle. During certain years the sun has large prominences on the surface that may be seen by looking at the sun through a piece of dark glass. These sunspots are of unknown origin. They are evidence of violent agitation on and within the sun. The occasional huge solar flare-ups or *sunspot storms* are apparently connected with sharp increases in ultraviolet radiation and result in the emission of electrified particles or *corpuscles* which bombard the atmosphere of the earth. The increase in radiation from the sun, as shown in Fig. 1-5, is responsible for *sudden ionospheric disturbances* (SIDs) while the corpuscular particles follow due to their slower speed. Fig. 1-6 is a graph that plots the present sunspot cycle, estimated on the basis of statistical probability as determined from previous cycles.

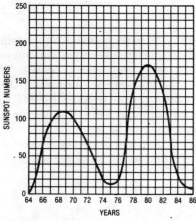

Fig. 1-6. Graph of sunspot frequency shows eleven-year cycle.

When conditions are right, the signal that otherwise would escape into space will come skipping back to earth at a distance far from the transmitter. There, the signal can rebound from the earth surface again. A series of such bounces between the earth and the ionized layers can take a signal around the globe. Each "bounce," however, robs a little bit of the signal strength. Thus, a high-powered signal from the other side of the world may be heard in North America at consistently good strength. And one from a lower-powered transmitter half the distance away may not have enough strength left after several skips to be heard clearly or even at all.

The ionosphere has several distinct layers as shown in Fig. 1-7. The layers are labeled D, E, and F. The latter has two segments, F1 and F2, but they merge into a single layer at night.

The D layer is closest to earth—roughly 50 miles high. It has the unfortunate habit of absorbing, rather than reflecting, the lower frequency shortwave and medium-wave signals. It is strongly affected by solar radiation, but quickly disappears after sunset. When the

D layer does vanish during the darkness hours, those radio signals can pass through to reach the reflecting ionized layers above. Perhaps, you have already noticed the daytime/nighttime phenomenon on the regular am medium-wave band. Cross-country stations, inaudible during the day, come skipping in during the night. The lower short-wave frequencies react the same way.

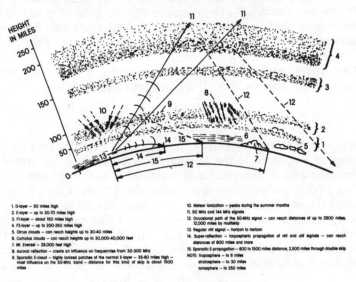

1. D-layer — 50 miles high
2. E-layer — up to 30-70 miles high
3. F1-layer — about 150 miles high
4. F2-layer — up to 200-350 miles high
5. Cirrus clouds — can reach heights up to 30-40 miles
6. Cumulus clouds — can reach heights up to 30,000-40,000 feet
7. Mt. Everest — 29,000 feet high
8. Auroral reflection — create an influence on frequencies from 30-300 MHz
9. Sporadic E-cloud — highly ionized patches of the normal E-layer — 35-60 miles high — most influence on the 50-MHz band — distance for this kind of skip is about 1500 miles

10. Meteor ionization — peaks during the summer months
11. 50 MHz and 144 MHz signals
12. Occasional path of the 50-MHz signal — can reach distances of up to 2500 miles, 12,000 miles by multiskip
13. Regular vhf signal — horizon to horizon
14. Super-reflection — tropospheric propagation of vhf and uhf signals — can reach distances of 800 miles and more
15. Sporadic E-propagation — 800 to 1500 miles distance, 2,500 miles through double skip
NOTE: troposphere — to 6 miles
        stratosphere — to 30 miles
        ionosphere — to 250 miles

**Fig. 1-7. Radio-wave propagation.**

Slightly higher in altitude is the E layer. It has some effect on the lower shortwave bands, but normally is of more interest to the vhf monitor or fm-tv DXers, since it accounts for such occasional long haul reception as occurs on these frequencies.

Several hundred miles up is the F layer of the ionosphere. It is the single most important factor in long distance shortwave reception. This is the belt of ionized gas that serves as the mirror to bounce most of the shortwave DX signals back for us to hear.

The shorter the wavelength of the signal, the easier it passes through the daytime D layer to reach the reflective layers above. So, over a daytime path between station and listener, the longer wavelengths— that is the higher frequencies—are the signals to tune in. Conversely, do not expect to hear a signal from around the world on, say 5,000 kHz over a fully sunlit path. The lower shortwave frequencies are nighttime DX bands.

This is the rule for daily propagation cycle: Use high frequencies during the day and lower frequencies at night.

# TROPOSPHERIC SKIP

Tropospheric skip is the most common form of vhf DXing. It is most pronounced in the vhf range and higher. It is the result of change in the reflection index of the atmosphere at the boundary between air masses of different temperature and humidity characteristics. These air masses often move on a very large scale, retaining their original character over considerable periods of time. A large mass of cold air is overrun by warm air and when this happens an inversion exists. This inversion produces good DXing in the vhf and uhf ranges.

This tropospheric bending is most common in fair, calm weather of the warmer months, though it can occur at any season.

# SPORADIC E SKIP

Sporadic E skip results from the reflection of vhf waves by dense patches of ionization in the E layer. Causes are still not completely understood and its occurrence is predictable only in a general way, but its effects are well known to many vhf DXers. Layer height and electron density determine the skip distance, but 50-MHz propagation is most common over distances of 400 to 1200 miles. Often signals are very strong, though they may vary rapidly over quite wide ranges. Ionization may develop simultaneously in several areas, making multiskip propagation possible over distances as great as 2500 miles.

# AURORAL PROPAGATION

Auroral propagation is the result of the sun emitting masses of hot, ionized gases, especially after solar flares. As these solar emissions flow toward the earth, they are split by the magnetic field of the earth and follow the magnetic lines to the other side of the earth where they concentrate around the magnetic poles. At the poles, the streams of energy from the sun strike the atmosphere at heights of 60 to 70 miles where they produce the vivid displays of the polar skies. Reflection of vhf signals from the auroral displays take place at a height of about 60 miles, where the propagation takes place in a generally east-west direction over a distance of several hundred miles —out to about 1200 miles on rare occasions.

# METEOR PROPAGATION

There are two types of meteors. The most common is very sporadic, arriving in view from random directions and traveling over a wide range of velocities. The second type appears when the earth sweeps through a meteor stream. These streams are believed to be the debris that follow a comet as it orbits the sun.

As a meteor disintegrates in its path across the sky, the process produces a burst of ionization around it. The ionization appears as a cylindrical column around the particle, expanding by diffusion through the rarified atmosphere. The ionized atmosphere is sufficiently large in many cases to affect a vhf signal momentarily. But the ionization produced by a stream of meteors is nearly continuous and it supports short bursts of vhf communication. It is these bursts—occasionally up to 40-seconds duration—that produce exceptional vhf DXing.

NOTE: The preceding paragraphs should give you some working knowledge of radio propagation and its effects on DX signals. However there is much more to radio signals than we talked about, but you can now haphazardly "scoot" about your radio dial and listen to the radio signals. The serious listener, however, will track the signals carefully. To do that you will have to learn to sort out the various numbers on your radio dial because on all modern receivers those numbers represent the frequencies.

## FREQUENCY BANDS

By international agreement, the radio spectrum is divided into various bands, of which the most popular to DXers are the shortwave international-broadcast bands and the radio-amateur bands. Other bands providing for good listening are the marine and nautical bands, the fm broadcast band, the "public-service" bands (reserved for police, fire, and municipal service), aeronautical and mobile communications, and space communications bands.

### Low Frequency

The low frequency (long-wave band) is 30 to 550 kHz (600 to 10,000 meters). This portion of the radio spectrum provides interesting reception for the listener who has learned to copy the International Morse code. It will provide a lot of fun and provide some interesting results. Stations that send cw (morse code) and radioteletype messages inhabit the lowest portion of the long-wave band. You can hear U.S. Navy stations in Washington and Maine, in England, Norway, Australia, and the time-signal station WWVB in Fort Collins, Colorado. There are also a number of other European, African, and Asian stations operating in this favorite segment of the long-wave band. Other stations that have been logged between 201 and 410 kHz include Morocco, Monaco, Poland, and Luxembourg. On the West Coast, the long-wave outlets of Radio Moscow in Siberia, Radio Peking, and Radio Mongolia have been heard. The long-wave broadcast band is for the "night owl," since it can best be heard during winter nights.

Do not be frightened by the thought of trying to copy a code station, because the stations transmitting cw code on the long-wave band make it easy. In addition to the very slow speed with which they send cw identifications, these stations often send "running markers" consisting of a series of V's followed by their call sign. With a little practice you can identify the code identification of the transmitting station sufficiently enough to make the sending of a reception report worthwhile.

It is altogether another story on the 410- to 500-kHz portion of the long-wave band, which is used by ship and coastal stations. You should have a fairly good knowledge of cw before tackling these stations since most of them operate cw at speeds of anywhere from 12 to 30 words per minute. Both ship and coastal stations operate on 500 kHz, one of the international distress and calling frequencies. A coastal station usually calls or answers ships on this frequency and then, if communications are to be continued, both the ship and coastal station move to a lower frequency, known as a "working frequency." Most freighter and tankers operate with 200- to 600-watts power output. Some passenger liners may use as much as 1,000 watts. These ship stations usually have a daytime range of 400 to 600 miles with a doubling or tripling of that range during the night.

Not many of the available commercial receivers, other than military surplus units, are equipped with a long-wave band. You will have to look, for example, for such units as the Zenith Transoceanic or the portable Grundig, Nordmende, and others.

## Medium Frequency

The medium frequency (medium-wave band) is 525 to 1605 kHz —(187 to 571 meters). This is called the "entertainment" band and is used world-wide for domestic broadcasting. American stations use 10-kHz spacing between channels, and that allows you to listen to the European stations, which have a totally different channel spacing. So, when you are on the East Coast you can listen to the powerful European senders and on the West Coast you can hear Australian, New Zealand, Japanese, Russian, or Chinese stations.

Broadcast band DXing—or BCB DXing, as it is called—is more difficult than shortwave DXing, because radio waves do not travel as well on broadcast band frequencies as they do on shortwave. Therefore, broadcast band stations, realizing that their coverage is limited, will aim their programs at local listeners, not at the international DXer—the shortwave DXer. As a BCB DXer you have a hard time hearing distant stations, and it is much easier to concentrate on the large number of U.S. stations.

**High Frequency**

The high frequency (shortwave band) is broken down into the following frequencies:

| kHz | MHz | Meter Band |
|---|---|---|
| 2300– 2495 | 2 | 120* |
| 3200– 3400 | 3 | 90* |
| 3900– 4000 | 4 | 75 |
| 4750– 5060 | 5 | 60* |
| 5950– 6200 | 6 | 49 |
| 7100– 7300 | 7 | 41 |
| 9500– 9775 | 9 | 31 |
| 11700–11975 | 11 | 25 |
| 15100–15450 | 15 | 19 |
| 17700–17900 | 17 | 16 |
| 21450–21750 | 21 | 13 |
| 25600–26100 | 26 | 11 |

(* = tropical bands, allocated to tropical countries)

The international shortwave broadcast stations are transmitting in the 13-, 16-, 19-, 25-, 31-, and 49-meter bands. You will find the tropical stations in the 60-, 90-, and 120-meter bands. And do not be surprised if you hear radioteletype stations mixed in with the tropical bands. A good thing to know is that the 13-, 16-, and 19-meter bands are susceptible to sunspot activities, while the 31-, 41-, and 49-meter bands should be clear year in and year out. Broadcasts in the high-frequency bands are capable of crossing frontiers at any time of the day or night, spanning oceans and bridging continents. For this reason a great amount of high-frequency broadcasting is devoted to international broadcasts—one country directly talking to listeners, like you, in another country thousands of miles away.

Reception conditions on the various high-frequency bands vary according to the time of day and season of the year.

Although not listed in our shortwave scale, the frequencies between 1605 and 2300 kHz are considered by many DXers to be part of the shortwave bands. During daylight hours, there is very little to be heard on these frequencies. At night, the 160-meter ham-radio band may be heard, long-range air navigation stations, a few South American airline beacons, a coastal station or two and a few ships might be picked up. For good DXing, therefore, concentrate your efforts on the SWBC bands listed in the remainder of this chapter.

Before we get into detail about the different shortwave bands, a few words about the tropical bands. As you can see in Table 1-1, the 60-, 90-, and 120-meter bands are allocated to the tropical countries —continents, actually—of Africa, Asia, Oceania, Central, and South

### Table 1-1. Tropical Band Allocations in kHz

| Location | Shortwave Band | | |
| --- | --- | --- | --- |
| | 120 meters | 90 meters | 60 meters |
| Africa, Asia, Oceania, South & Central America | 2300–2500 | 3200–3400 | 4750–5060 |

America. (The 75-meter band is not considered a tropical band in the western hemisphere.) In Europe and the North American continent, free enterprise and governmental broadcast authorities have enough money and manpower to construct a sufficient number of am and fm stations in order to give everyone a reasonably good variety of programs from which to choose. This is not the case in the so-called "third world." Financially, they are not capable of installing a vast number of transmitters and studios. Therefore, insurmountable problems are associated with covering their nations with a blanket of am and fm stations. So, in many countries shortwave radio is the only contact people have with the "outside world." These shortwave broadcasting bands—over the region of 2000 to 6000 kHz—are ideal for the purpose of educational and entertainment programs for their citizens. These radio signals are not deterred by natural obstacles and can penetrate into the most isolated areas. So, these bands are used by infinitely varied stations in the most exotic lands of the world, such as Malaysia, Indonesia, Solomon Islands, Peru, Brazil, and many more. These stations are much like the U.S. am and fm stations; they have disc jockeys, hit records, classical music, news, commercials, sports, etc. The only difference is that these radio broadcasts take place in the shortwave bands.

### 120-Meter Band—2,300–2,495 kHz

There are not too many stations that operate in this band, but those that do can generally be considered good DXing. Daylight reception of this band is never possible since the D-layer continues to exert its influence right up until sunset. Even at night conditions are often not good enough to propagate the low-powered signals used by the broadcast stations. Keep trying, however, because an opening might bring in a broadcaster.

### 90-Meter Band—3,200–3,400 kHz

This is the "night" band. From dusk through dawn, with the exception of the middle-of-the-night period when they go off the air, you can find a number of Central and South American stations. During the late night hours there will be some African stations coming on the air. It is early morning on the other side of the Atlantic, and the broadcasters are then signing on to start their daily schedule. Also, from shortly before your local dawn until just after the sun rises, Asian and Pacific stations filter through.

Although good conditions are not unusual on this band, a variety of interference must be expected. Summer thunderstorm interference, radioteletype stations, and other interferences that have increased in the past few years may create certain annoyances.

### 60-Meter Band—4,750–5,060 kHz

This is the band with most of the activity—again, primarily at night. It is on this band that you will find the local and regional home-service outlets of the tropical countries . . . more than seventy-five of them. It can take months before you sort out the various stations that can be heard. But no other band can match the excitement and thrill of listening to local programs from around the world. DXing this band will give you a glimpse of the tempo of life in those faraway lands. Proficiency in DXing these bands depends on your experience, but learning all of the "ins-and-outs" of this tropical band is quite enjoyable. Do not rush, however. Be patient, especially when interferences are encountered. Noise, hash, lightning static, and the transmissions of utility stations may hamper your DXing efforts. Keep trying and eventually the wildest of your DXing dreams may be fulfilled.

### 75-Meter Band—3,900–4,000 kHz

This is not considered a broadcast band in the western hemisphere, but it is in other parts of the world. It is plagued with heavy interference because it is shared with the ham operators in the U.S. and Canada. However, certain signals are sometimes quite audible, especially above 3,950 kHz, which is allocated for European use in international broadcasting. Some better DX possibilities include the Cape Verde Islands, Greenland, Lagos (in Nigeria), South Africa, Indonesian local-service stations, and other parts of the Pacific.

### 49-Meter Band—5,950–6,200 kHz

This is a very crowded band because it is used by both the local-service stations and international broadcasting stations. The BBC in England and the Deutsche Welle in West Germany, for example, share this band with African and South American local stations. Australian and other Asian signals can be heard as well. This band opens up earlier in the afternoon and closes somewhat later in the morning, but it is primarily a night band.

### 41-Meter Band—7,100–7,300 kHz

Like the 75-meter band, this band is shared between the international broadcasting stations and the ham operators, which is not too ideal a situation. By international agreement, this sharing of these particular frequencies is permitted, but as a DXer trying to listen to the international stations, you will have a problem, especially when one ham station is trying to work another a mile away and a broadcast station with 100,000 watts of power opens up on the same frequency. You will have the same problem when you are trying to monitor an international station with less power.

### 31-Meter Band—9,500–9,775 kHz

This "middle" band may well be the best range of frequencies when you are a beginning DXer. Many of the large international broadcasting stations operate on this band, such as stations with powerful transmitters in Europe, Africa, and other parts of the world. These are good for listening most any time of the night and during daylight hours, too.

### 25-Meter Band—11,700–11,975 kHz

This band has some of the characteristics of the 31-meter band. The same kind of international broadcasters use this band rather extensively. It is an excellent band for DXing at almost any time, but it is crowded during the early afternoon and evening.

### 19-Meter Band—15,100–15,450 kHz

This is a good band to tune in during daylight hours—the signals die out in the evening. You can DX for a number of powerful European stations and some mideastern stations. Early evenings you may be able to pick up some South American broadcasters and a favorite station among many DXers: Radio Tahiti.

### 16-Meter Band—17,700–17,900 kHz

This band is an excellent one for the DXer who has a lot of time during the daytime hours—the time to tune in and listen to the large international broadcasters.

### 13-Meter Band—21,450–21,750 kHz

The story is much the same here as it is with the 16-meter band. There is a lot of daytime activity with mostly international broadcasters, but with less activity than there was a number of years ago. There are, however, about 60 stations to listen to, including broadcasters from Belgium, Morocco, Madagascar, Norway, the Middle East, the Philippines, the Soviet Union, and many others.

### 11-Meter Band—25,600–26,100 kHz

This band is the highest of the 12 shortwave bands, and there is not much activity to be heard on these frequencies.

**Other Frequencies**

The bulk of the shortwave broadcast signals are to be found in these 12 SWBC bands. What about the "spaces" or "holes" between the bands? The area around the 2,400–2,600 kHz, for example, is used for ship-to-shore voice transmissions. Most of the time you will hear the coastal stations only, since many of the ships are operating a bit lower in frequency. At times, however, you actually can hear both sides of the conversation. The 2,600–2,700 kHz range is used for voice communications from ships on inland waterways.

There are also a number of stations between 5,100 and 5,900 kHz. Between 6,200 and 7,100 kHz, you may again hear some shortwave broadcast stations that are apparently operating outside of the recognized shortwave broadcast bands. Try listening and, if you are lucky, you may find yourself listening to China, North Korea, Inner Mongolia, Vietnam, or Pakistan.

Radio Peking in China uses the frequency range between 7,400 and 9,000 kHz for many of its services. You may also hear voice stations from airline terminals and planes, coastal stations in the ship-to-shore service for long-distance Morse communications, radio-teletype stations, and a few military tactical stations, all of which are operating in Morse code.

From 9,800 to 11,700 kHz, there are, again, a good many broadcasting stations that operate outside of the recognized shortwave bands. Most of these stations are located in countries such as China, Russia, and North Vietnam.

The frequencies between 12,000 and 15,000 kHz are used largely by stations using Morse code for a wide variety of communications. Ham radio operators can be heard between 14,000 and 14,350 kHz—the favorite long-distance band for amateur operators. A complete list of the amateur bands is presented in Table 1-2.

Amateurs also communicate by means of two satellites: Oscar 6 and Oscar 7. Oscar 6 transmits down to earth between 29.45 and 29.55 MHz. Oscar 7 transmits down on 29.4–29.5 MHz.

**Table 1-2. Amateur-Radio Bands**

| Frequency kHz | Wavelength Meters | Frequency MHz | Wavelength Meters | Frequency GHz | Wavelength Meters |
|---|---|---|---|---|---|
| 1800–2000 | 160 | 50–54 | 6 | 10.0–10.5 | — |
| 3500–4000 | 75/80 | 144–146 | 2 | 24.0–24.25 | — |
| 7000–7300 | 40 | 220–225 | 1¼ | 48–50 | — |
| 14000–14350 | 20 | 420–450 | ¾ | 71–76 | — |
| 21000–22450 | 15 | 1215–1300 | — | 165–170 | — |
| 28000–29700 | 10 | 2300–2450 | — | Above 300 | — |
|  |  | 3300–3500 | — |  |  |
|  |  | 5650–5925 | — |  |  |

Many of the satellites do not transmit continually, but are programmed to transmit on receiving a trigger signal from the ground. Table 1-3 lists the assigned satellite frequencies.

### Table 1-3. Assigned Communication Satellite Frequencies

| Frequency MHz | Name | Nation | Frequency MHz | Name | Nation |
|---|---|---|---|---|---|
| 40.01 | Intasat | Spain | 136.89 | Explorer 47 | U.S. |
| 41.01 | Intasat | Spain | 136.92 | OSO 8 | U.S. |
| 136.08 | ISIS 1 | Canada | 136.95 | Nimbus 4 | U.S. |
| 136.23 | ATS 6 | U.S. | | COS 8 | U.S.S.R. |
| 136.25 | Castor | France | 137.11 | ATS 6 | U.S. |
| 136.26 | Copernicus | U.S. | 137.14 | NOAA 3 | U.S. |
| 136.29 | Explorer 52 | U.S. | | NOAA 4 | U.S. |
| 136.32 | Explorer 32 | U.S. | 137.23 | Explorer 51 | U.S. |
| 136.32 | Geos 2 | U.S. | | Explorer 54 | U.S. |
| 136.38 | OV5-6 | U.S. | 137.35 | ATS 1 | U.S. |
| 136.38 | Explorer 44 | U.S. | | ATS 3 | U.S. |
| 136.38 | Intercosmos 2 | U.S. | | ATS 5 | U.S. |
| 136.38 | SMS 2 | U.S. | 137.50 | NOAA 3 | U.S. |
| 136.38 | GOES 1 | U.S. | | NOAA 4 | U.S. |
| 136.41 | ISIS 1 | Canada | 137.53 | SRET 2 | France |
| 136.41 | ISIS 2 | Canada | 137.62 | ESSA 8 | U.S. |
| 136.44 | Copernicus | U.S. | | NOAA 3 | U.S. |
| 136.47 | ATS 1 | U.S. | | NOAA 4 | U.S. |
| | ATS 3 | U.S. | 137.68 | ARIEL 5 | U.S.S.R. |
| | ATS 5 | U.S. | 137.86 | Landsat 1 | U.S. |
| 136.50 | Nimbus 4 | U.S. | | Landsat 2 | U.S. |
| | Nimbus 5 | U.S. | 137.89 | ANS | The Netherlands |
| | Nimbus 6 | U.S. | 137.92 | Explorer 47 | U.S. |
| 136.59 | ISIS 1 | Canada | 137.95 | ISIS 1 | Canada |
| | ISIS 2 | Canada | | ISIS 2 | Canada |
| 136.68 | Explorer 53 | U.S. | 137.98 | Explorer 50 | U.S. |
| 136.71 | Intasat | Spain | 400.55 | Copernicus | U.S. |
| 136.74 | Explorer 27 | U.S. | 400.65 | Explorer 52 | U.S. |
| | San Marco 4 | Italy | 400.95 | Explorer 49 | U.S. |
| | D2-8 | France | 401.20 | Nimbus 6 | U.S. |
| 136.77 | ESSA 8 | U.S. | 401.50 | Nimbus 4 | U.S. |
| | NOAA 3 | U.S. | 401.75 | ISIS 1 | Canada |
| | NOAA 4 | U.S. | 468.825 | Intercosmos 2 | U.S.S.R. |
| 136.80 | Explorer 50 | U.S. | | SMS 2 | U.S. |
| 136.86 | Explorer 49 | U.S. | | GOES 1 | U.S. |

The various Morse code services can be heard again in the frequency range of 15,500 to 17,700 kHz. There are also a few voice broadcasters here and there.

Why do some countries operate outside the recognized bands? That is difficult to say. Ignorance, perhaps. Or questionable technical standards? The most logical answer could be, "the fight for space." Because it is always easier to find a clear channel in the "holes."

# FOREIGN BROADCAST BAND FREQUENCIES

Under proper conditions, all areas of the United States have a chance of picking up signals from the West Coast of Africa, Australia, and east Asia. Most BCB DXers (Broadcast Band) confine their listening to the autumn and winter months, typically from early September to early April. See Table 1-4 for a list of foreign broadcast band frequencies.

## Table 1-4. Foreign Broadcast Band Frequencies

| Latin America and Caribbean Frequencies | | | |
|---|---|---|---|
| kHz | Country | kHz | Country |
| 540 | Mexico | 825 | Costa Rica |
| 550 | Cuba | 834 | Belize |
| 595 | Dominican Rep. | 900 | Mexico |
| 600 | Cuba | 1035 | Haiti |
| 640 | Cuba | 1055 | Colombia |
| 655 | El Salvador | 1200 | Venezuela |
| 675 | Costa Rica | 1265 | St. Kitts |
| 725 | Surinam | 1555 | Cayman Islands |
| 800 | Netherl. Antilles | 1570 | Mexico |

| Transpacific Frequencies | | | |
|---|---|---|---|
| kHz | Country | kHz | Country |
| 655 | North Korea | 844 | Gilbert Islands |
| 750 | Japan | 877 | North Korea |
| 770 | Japan | 1040 | China |
| 830 | Japan | 1525 | China |
| 835 | China | 1550 | Australia |

| Transatlantic Frequencies | | | |
|---|---|---|---|
| kHz | Country | kHz | Country |
| 665 | Portugal | 1394 | Albania |
| 737 | Spain | 1403 | Guinea |
| 764 | Senegal | 1466 | Monaco |
| 845 | Italy | 1475 | Austria |
| 854 | Spain | 1538 | West Germany |
| 1016 | West Germany | 1554 | France |
| 1016 | Turkey | 1562 | Switzerland |
| 1205 | France | 1586 | West Germany |
| 1214 | England | | |

If you are tuning in for stations in Latin America or the Caribbean, you should tune in from shortly after your local sunset until sunrise. DXers in the eastern United States should tune in for Europe and Africa before 0000 GMT—reception will last until approximately 0700 GMT, when sunrise in Europe and Africa ends the chances for reception.

Tuning in to Oceania and Asia should be done from 0700 GMT onward during the autumn/winter DX season. It will remain possible until roughly 1300 GMT.

Where can more information be found about DX activities? Chapter 5 lists all the DX clubs—domestic and international—which you can join by paying a low yearly contribution. These clubs will keep you informed about a variety of DX activities around the world.

## VERY HIGH AND ULTRA HIGH FREQUENCY BANDS

A breakdown of the higher frequency bands by their assigned frequencies is as follows:

Very High Frequency (vhf)—30,000–300,000 kHz or 30–300 MHz.

### Chart 1-1. Broadcast Media and Frequency Allocations

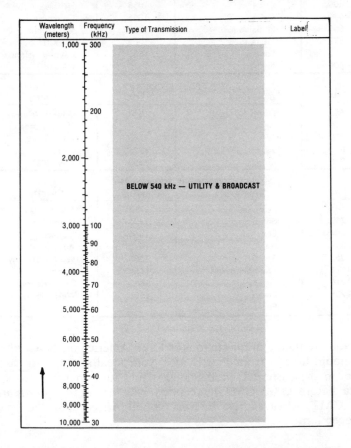

## Chart 1-1 (Cont). Broadcast Media and Frequency Allocations

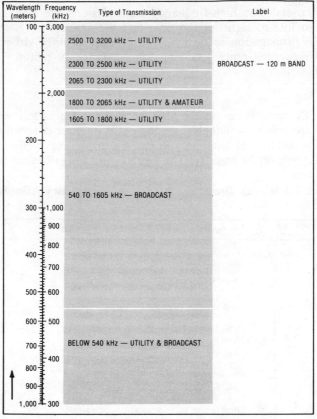

| Wavelength (meters) / Frequency (kHz) | Type of Transmission | Label |
|---|---|---|
| | 2500 TO 3200 kHz — UTILITY | |
| | 2300 TO 2500 kHz — UTILITY | BROADCAST — 120 m BAND |
| | 2065 TO 2300 kHz — UTILITY | |
| | 1800 TO 2065 kHz — UTILITY & AMATEUR | |
| | 1605 TO 1800 kHz — UTILITY | |
| | 540 TO 1605 kHz — BROADCAST | |
| | BELOW 540 kHz — UTILITY & BROADCAST | |

Ultra High Frequency (uhf)—300–3000 MHz.
Super High Frequency (shf)—3–30 GHz.
Extremely High Frequency (ehf)—30–300 GHz.

There is a whole new world of DXing in the frequency spectrum above 30 MHz. All tv broadcasting takes place between 54 and 890 MHz. The fm broadcast stations are located between 88 and 108 MHz. Police and fire departments, the forestry services, taxis, construction companies, and other businesses occupy the so-called "Public Service" bands between 30 and 50 MHz, 148 and 174 MHz, and above 450 MHz. Interspersed between and above these frequencies are amateur bands, government and military channels, radar services, and weather and other satellites.

The normal range of the vhf and uhf signals is less than 100 miles —their common denominator is reliability over short distances, freedom from static, and the vast amount of spectrum space within them.

This limited radio range permits the duplication of frequency assignments every few hundred miles across the country without interference between users. The high-frequency transmissions are not, however, limited to line of sight all the time and the DXer who is aware of the radio propagation mode involved in long distance vhf reception can have exciting times receiving long distance tv pictures and fm stations.

## TV-DXing

The DX possibilities on the high frequencies are as follows:
*DX tv reception*—channels 2 through 6. With your tv receiver and a high gain, directional antenna you can try to pick up tv signals from different parts of the country. The key to good long distance tv recep-

**Chart 1-1 (Cont). Broadcast Media and Frequency Allocations**

| Wavelength (meters) | Frequency (kHz) | Type of Transmission | Label |
|---|---|---|---|
| 10 | 30,000 | 29700 TO 30000 kHz — UTILITY | |
| | | 28000 TO 29700 kHz — AMATEUR | 10 m BAND |
| | | 27230 TO 28000 kHz — UTILITY | |
| | | 26960 TO 27230 kHz — AMATEUR - CB | |
| | | 26100 TO 26960 kHz — UTILITY | |
| | | 25600 TO 26100 kHz — BROADCAST | 11 m BAND |
| | | 21750 TO 25600 kHz — UTILITY | |
| | | 21450 TO 21750 — BROADCAST | 13 m BAND |
| | | 21000 TO 21450 — AMATEUR | 15 m BAND |
| | 20,000 | 17900 TO 21000 kHz — UTILITY | |
| | | 17700 TO 17900 kHz — BROADCAST | 16 m BAND |
| | | 15450 TO 17700 kHz — UTILITY | |
| 20 | | 15100 TO 15450 kHz — BROADCAST | 19 m BAND |
| | | 14350 TO 15100 kHz — UTILITY | |
| | | 14000 TO 14350 kHz — AMATEUR | 20 m BAND |
| | | 11975 TO 14000 kHz — UTILITY | |
| | | 11700 TO 11975 kHz — BROADCAST | 25 m BAND |
| | | 10000 TO 11700 kHz — UTILITY | |
| 30 | 10,000 | 9775 TO 10000 kHz — UTILITY - BROADCAST | |
| | | 9500 TO 9775 kHz — BROADCAST | 31 m BAND |
| | 9,000 | | |
| | 8,000 | 7300 TO 9500 kHz — UTILITY | |
| 40 | 7,000 | 7100 TO 7300 kHz — AMATEUR - BROADCAST | 41 m BAND |
| | | 7000 TO 7100 kHz — AMATEUR | |
| | | 6200 TO 7000 kHz — UTILITY | |
| 50 | 6,000 | 5950 TO 6200 kHz — BROADCAST | 49 m BAND |
| | | 5060 TO 5950 kHz — UTILITY | |
| 60 | 5,000 | 4750 TO 5060 kHz — BROADCAST | 60 m BAND |
| 70 | | 4000 TO 4750 kHz — UTILITY | |
| | 4,000 | 3900 TO 4000 kHz — UTILITY - AMATEUR - BROADCAST | 75 m BAND |
| 80 | | 3500 TO 3900 kHz — UTILITY - AMATEUR | 80 m BAND |
| 90 | | 3400 TO 3500 kHz — UTILITY | |
| | | 3200 TO 3400 kHz — UTILITY | |
| 100 | 3,000 | 2500 TO 3200 kHz — UTILITY BROADCAST 90 m BAND | |

## Chart 1-1 (Cont). Broadcast Media and Frequency Allocations

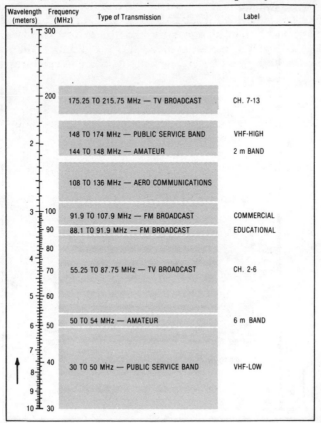

| Wavelength (meters) | Frequency (MHz) | Type of Transmission | Label |
|---|---|---|---|
| 1 | 300 | | |
| | 200 | 175.25 TO 215.75 MHz — TV BROADCAST | CH. 7-13 |
| 2 | | 148 TO 174 MHz — PUBLIC SERVICE BAND | VHF-HIGH |
| | | 144 TO 148 MHz — AMATEUR | 2 m BAND |
| | | 108 TO 136 MHz — AERO COMMUNICATIONS | |
| 3 | 100 | 91.9 TO 107.9 MHz — FM BROADCAST | COMMERCIAL |
| | 90 | 88.1 TO 91.9 MHz — FM BROADCAST | EDUCATIONAL |
| 4 | 80 | | |
| | 70 | 55.25 TO 87.75 MHz — TV BROADCAST | CH. 2-6 |
| 5 | 60 | | |
| 6 | 50 | 50 TO 54 MHz — AMATEUR | 6 m BAND |
| 7 | | | |
| 8 | 40 | 30 TO 50 MHz — PUBLIC SERVICE BAND | VHF-LOW |
| 9 | | | |
| 10 | 30 | | |

tion is your antenna. Also an antenna rotator is a must—the kind that provides continuous rotation. An antenna preamplifier is a great help, because it builds up the weak signals so that they can be better identified.

For uhf reception, a parabolic disc is hard to beat—you need at least a 5- or 7-foot disc.

Sporadic E skip—See "Propagation" in this chapter—builds up from the lower frequencies, reaching the lower tv channels and occasionally rising as high as channel 6. Reception of the lower tv channels during a sporadic E propagation opening provides reception up to about 1,000 miles. Reception of sporadic E signals below 500 miles is rare, and the upper distance for this type of skip is about 1,500 miles.

On some occasions, double-hop sporadic E propagation may be observed, caused by two ionized patches between the DXer and the station. This permits reception of tv signals up to 2,500 miles away.

## Chart 1-1 (Cont). Broadcast Media and Frequency Allocations

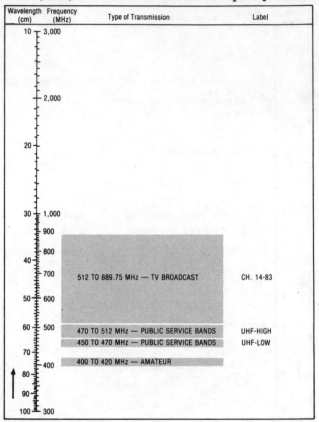

| Wavelength (cm) | Frequency (MHz) | Type of Transmission | Label |
|---|---|---|---|
| | | 512 TO 889.75 MHz — TV BROADCAST | CH. 14-83 |
| | | 470 TO 512 MHz — PUBLIC SERVICE BANDS | UHF-HIGH |
| | | 450 TO 470 MHz — PUBLIC SERVICE BANDS | UHF-LOW |
| | | 400 TO 420 MHz — AMATEUR | |

A double skip is much more likely to occur on channel 2 than on the higher channels.

The best way to get started in sporadic E long distance reception is to leave your tv set tuned to channel 2 or 3 and watch for a picture during the months of maximum E propagation—the north-south path of which seems to be more prevalent during the summer months.

Temperature inversion, which also affects tv DXing, forms along a stationary air front where a cool high pressure system meets warmer air. At the junction of these air masses an inversion can extend for 800 miles or more. Instances of tv reception between the U.S. mainland and Hawaii have occurred by a form of temperature inversion that provides an atmospheric duct which acts much in the manner of a *radio hose,* propagating the vhf signals over distances in excess of 2,500 miles. Check weather reports and weather maps; they will help you to spot weather conditions favorable for extended *tropo-*

*spheric* tv-DX reception. Look especially for slow moving highs followed by cold fronts, such as those shown in Fig. 1-8. Fog, smog, and haze are also indications of temperature inversions.

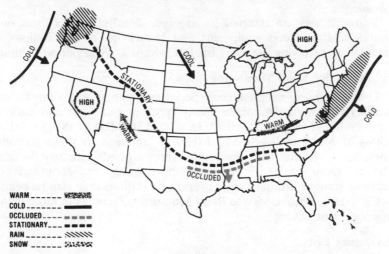

WARM _____
COLD _____
OCCLUDED ____
STATIONARY ___
RAIN _____
SNOW _____

**Fig. 1-8. Check for weather conditions favorable for extended tropospheric tv-DX reception.**

Tropospheric reception is best on the higher tv channels and on the uhf channels; it is poorest on the lower channels. Sometimes during the fall and spring, tv stations from the tropics blanket the entire Gulf Coast as far inland as 300 miles.

During periods of high sunspot activity, spectacular tv DXing is possible with F2 propagation (covered in this chapter under "Propagation"). While the sunspot cycle will not reach a new peak until about 1980, short periods of erratic F2 skip are possible during spring and fall months.

### FM DXing

The fm radio band is assigned a most ideal portion of the electromagnetic spectrum, since it benefits from all kinds of propagation. Within its 88 to 108 MHz confines, you can experience reception by means of skip, extended ground wave, tropo and line-of-sight reception of local stations—all with beautiful high-fidelity monophonic sound, or in full stereo. Adding to the excitement of fm DXing is the reception of weak signals by meteor scatter, auroral propagation, etc.

In the U.S. and Canada the fm band contains 100 DX-ready very high frequency channels, comparable in their reception characteristics to tv channels 2 through 6, with the fm band lying immediately above channel 6 in the spectrum.

Equipment for fm DXing need not be elaborate—especially for skip, which occurs primarily during early and mid-summer, mainly during the day and early evening. A good fm receiver will do the job adequately.

You also need an external fm antenna. Ideally, a directional one with a rotator. Good equipment and knowledge of the stations is essential to receiving the most fm DX within a given period of time.

### Public Service Band Monitoring

Police, highway maintenance, special emergency, and aeronautical stations are only a few of the many types of radio communications services found in the 30–50 MHz, 108–136 MHz, 148–174 MHz, 450–470 MHz, and 470–512 MHz. In addition to these stations, normally associated with the public service bands, are mobile telephones, radio pagers, broadcast remote links, and thousands of business stations. Military and government stations may also be heard.

Chapter 6, "Public Service Band Monitoring," covers in more detail this aspect of DXing.

### Frequency Lists

While a shortwave frequency list (Chapter 5 contains two such references) is a must for every DXer—the DXer who wants to listen to broadcasts that are exciting and unique will find the "Confidential Frequency List" indispensable. Published by Gilfer Associates, Inc., 52 Park Ave., Park Ridge, NJ 07656 ($5.45), this book lists such frequencies as: Foreign and U.S. military; Interpol; Communist Broadcast Feeders; U.S. Border Patrol; Embassy Stations, etc.

Of course, the DXer is reminded that "the U.S. Communications Act of 1934, Section 305, as amended (47 USC, Section 605) prohibits anyone, not authorized by the sender, from intercepting any radio communication not intended for reception by the general public, and divulging or beneficially using that information from such monitoring." In other words, listen by all means, but do not benefit from or divulge the information obtained.

Now that you are familiar with the different bands on which to listen to the "sounds of the world around you," you are wondering, "How do I buy a receiver and what do I buy?" The next chapter will help you answer these questions.

## DXING THE BANDS

There are three very broad categories of stations you can DX. First, there are the broadcast stations which air programs of information, entertainment, and music to general audiences. Second, there are the radio amateurs, hobbyists who operate on the shortwave and higher frequencies. And third, everything else falls into the general category—utilities, which cover the whole range of services from longwave ship-to-shore to aeronautical signals, and point-to-point transatlantic phone calls to radio teleprinters, etc.

The frequency charts provided on the following pages give you a complete overview as to where these various stations are located in the broadcast-frequency spectrum. Chart 1-1 is presented in four columns, giving such information as:

*Wavelength:* from 10,000 meters to 1 meter, and from 100 centimeters to 10 centimeters

*Equivalent frequencies:* from 30 kHz (0.03 MHz) to 30,000 kHz (30 MHz) and from 30 MHz to 3,000 MHz

*Type of transmission:* transmitting that takes place in a certain frequency spectrum—broadcast, amateur, tv, etc.

*Label:* category given to a number of frequencies close together in the spectrum, e.g., 120-m band, 10-m band, etc.

The chart will help you to know what kind of broadcast media to expect from different frequency allocations.

Fig. 1-9 shows the wavelengths and frequencies on which you can receive amateur radio—ham—communications.

These figures will make it easier for you in selecting the proper equipment based on your particular desires of listening to certain frequencies/stations.

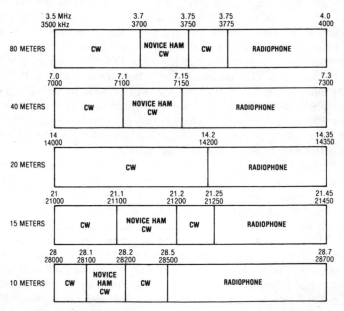

**Fig. 1-9. Ham radio bands and reception frequencies.**

# CHAPTER 2

# How To Shop for DX Receivers

What is desired versus what is required is the problem you will run into as a new DXer. There is no "best" receiver—there are dozens available with various features to aid you in DXing. It will be up to you to make up your own mind and to find that one solution for your specific situation. The following are some general tips to help you reach that decision.

A beginning listener may not have to invest a dime to get started in DXing. The BCBer (Broadcast Band DXer) can start with almost any radio that covers the am medium-wave band. For shortwave listening it might be a multiband portable you already own.

But whatever receiver you use for your DXing, learn to use it, tax it to its maximum DX potential, and then trade up to a better set. At this time, it is more likely that you will be able to sell or trade the old set and move up without making an additional investment that is too great. And, by the time you get that set you have been dreaming about, you will have built up your DX skills to the point where you can use the new radio to the best advantage.

Price is one factor that is critical. A new communications receiver does not sell for pennies. Yet, DXing is not an expensive hobby. There are some very good receivers on the market today—not as wide a selection as there was a few years back, perhaps, but those sold today are fine DX radios. It is doubtful that you will hear anything on a $3,000 receiver that you would not hear on a $300 set, if you are sufficiently skilled and have patience. The fancy high-priced unit may make it easier, may make real DX loggings more regular, or simply make listening more fun. But do not let the fact that you have only a few hundred dollars to spend dishearten you. DXing can be a great pastime even if you have only the simplest gear.

All kinds of varieties are possible, depending on the use you will make from it and the price you are willing to pay. The most important property to judge whether a receiver is specially designed for your needs is the frequency coverage. Not only the long waves—150 to 400 kHz—and the medium waves—500 to 1600 kHz—but also the shortwaves from 1.6 MHz to at least 25 or 30 MHz should be on the scale of the receiver. And the shortwave coverage should be divided in at least four or five different parts, otherwise it will be completely impossible for you to tune in on a certain frequency that you like or to read the scale at the frequency you are actually listening to. The more tuning ranges the receiver has, the better and the more accurate it will be.

But the frequency coverage is not the only important property of a receiver. A good set should at least have the following features.

## SENSITIVITY

Any quality receiver today usually has an rf stage, and you can look for this in the specifications. The rf stage helps achieve the second criterion: good signal-to-noise ratio. Signal-to-noise ratio (S/N) is usually printed in the specifications of the receiver. Typically, a good receiver will have a sensitivity of 1 microvolt in order to achieve a 10-dB signal-to-noise ratio. This is written as "1 microvolt for 10 dB S/N." Some receivers will have less than 1-microvolt sensitivity. The lower the figure the better the set. There is a quick and easy test you can make, even if you are in the sales showroom. If the dealer has an antenna available, tune across all the bands with the antenna connected. Do not be concerned with the stations you can hear; listen, instead, for the background noise between stations. Next, disconnect the antenna and make the same check. A set without the antenna connected should produce significantly less background noise. The noises you hear with the antenna disconnected are produced by the receiver circuitry.

## SELECTIVITY

Selectivity means being able to bring in the wanted station but rejecting all, or most, other signals on adjoining frequencies—in short, rejecting interference. Today's shortwave bands are crowded and it takes a good receiver to sort out the desired signal from the unwanted ones. Without adequate selectivity, your receiver will create a bedlam of noise and unintelligibility. Your communications receiver must have very good selectivity—much better than the usual home radio.

Selectivity may be achieved in a number of ways or, more commonly, in a combination of ways. A superheterodyne receiver makes use of an intermediate-frequency amplifier—the so-called if stage—to produce basic selectivity. A double or triple conversion model with multiple if stages will provide better selectivity—but, it costs more!

## STABILITY

Stability is the ability of the receiver to resist drifting after warm-up and to remain on frequency even after band switching. Tune to a station and check whether it is still there fifteen minutes later. There are two different things that affect stability. One is the ability of the receiver to compensate for temperature changes within the set as tubes (in older sets) and various other components warm up. The other is the mechanical stability of parts and the cabinet. One of the advantages of the newer solid-state—transistorized—receivers is that they show less tendency to drift.

A number of different techniques are used to achieve electronic stability, including the utilization of temperature-compensating capacitors and a crystal-controlled–oscillator stage. The data supplied in the advertising brochures describing a good receiver will usually indicate the number of hertz of drift during warmup. A quality receiver, for example, will drift only 400 Hz from a cold start.

One practical test of stability you can apply to a receiver is to tune it to a long-winded ssb (single-sideband) phone station on as high a frequency as possible. Check 15 minutes later and see how much retuning is necessary to keep the signal in tune and intelligible. Another test is to tune in to a code (cw) station or am station with the bfo (beat frequency oscillator) ON, then to lift the set by one corner off the table and to drop it an inch or two. A good receiver will take a reasonable amount of this kind of treatment without serious detuning.

## SIGNAL-STRENGTH INDICATOR

The signal-strength meter serves two purposes. First, it indicates the signal strength of the station you are tuned in to and second, it tells you the moment your receiver is properly tuned in.

## NOISE LIMITER

A noise limiter helps to eliminate noise interference, such as ignition noises and interference from electrical appliances. Some sets have a front-panel noise-limiter control to allow you the choice of noise limiting. Signal strength is reduced when the noise limiter is turned ON.

## AUTOMATIC GAIN CONTROL

A number of receivers utilize front panel automatic gain control (agc) to allow you to achieve maximum amplification when needed —so you can "pull-in" those "weak, far-away" stations.

## BEAT-FREQUENCY OSCILLATOR—BFO

This is a device from which an audible signal is obtained by combining and rectifying two higher inaudible frequencies. Especially needed to receive ssb and cw—Morse code—transmissions. Without a bfo, ssb transmissions received with your set sound like Donald Duck talking.

## CALIBRATOR

The purpose of the calibrator is to allow you to set your tuning dial so that when the dial indicates a certain frequency, you are actually close to that frequency. This is of great help not only for locating the band in the first place, but also for indicating where the edges of the band are located.

More expensive receivers usually feature an electronic digital readout. If you want to tune to a certain frequency, you simply do so until your illuminated numeral readout indicates you are tuned in to that frequency. You will obtain tuning accuracy to a fraction of 1 kHz. There are also sets available with highly accurate mechanical-type readouts of frequency—at least to the nearest kilohertz, and with interpolation to fractional kilohertz.

The less expensive sets have a slide-rule–type system in which a bar marker moves across a marked dial plate. With these sets, it is often hard to tell whether you are tuned to 9,600 or 9,650 kHz. However, this may not be important to you. If you are out "hunting" for stations, a digital readout will be handy, but not necessary. And you can always return to a certain station by making a notation of your dial setting.

## A HELPFUL CHECK LIST

The following check list will provide you with a convenient aid when considering a specific receiver.

Sensitivity: _____ microvolt for 10 dB S/N (1.5 $\mu$V or less).
Selectivity: _____ kHz at 6 dB (less than 5 kHz is ideal).
Stability: _____ Hz (100 Hz or less is ideal).
Signal Strength Indicator

Noise Limiter Control
Automatic Gain Control
BFO—Front Panel Control
Calibration Accuracy: ———— kHz (1 kHz is ideal).
Hum & Noise: ———— dB (40 to 60 dB is ideal).

## A SHOPPING GUIDE

Figs. 2-1 through 2-4 show a variety of DX receivers available on the market. They range from $350 for the SSR-1 receiver with built-in

**Fig. 2-1. SSR-1 shortwave receiver covers 0.5 to 30 MHz and operates on 12 vdc or 117/230 vac.**

**Fig. 2-2. DR22 receiver with digital readout operates on 117 vac.**

telescoping antenna, shown in Fig. 2-1, to the more expensive—about $3,000—HRO 500 with four built-in bandwidth selections: 500; 2,500; 5,000; and 8,000 Hz and tuning from 0.5 to 30 MHz in sixty 500-kHz–wide segments shown in Fig. 2-3. A general coverage receiver, the DR22, is shown in Fig. 2-2. This model covers 0.5 to

Fig. 2-3. HRO 500 communications receiver operates on 12 vdc or
117/230 vac.

Fig. 2-4. Galaxy Mesa 6606 portable receiver covers nine
communications bands.

29.7 MHz, offers digital readout, and costs approximately $1,000.
Portable receivers like the Galaxy Mesa 6606 in Fig. 2-4 range in
price from $200 to $2,000. This model covers nine bands, including
49-m, 41-m, 31-m, 25-m, 19-m, and 16-m bands; am, fm, and long

wave (145 to 260 kHz) and operates on six "C" batteries or 117 vac. It has a built-in telescoping antenna.

Communications receivers are available from several manufacturers and suppliers which include:

| | |
|---|---|
| Collins | National Radio |
| Drake | Normende/Galaxy |
| Dames | Panasonic |
| Dymek | Radio Shack |
| General Electric | Sony |
| Kenwood | Yaesu-Musen |
| Heathkit | Zenith |

Communications receivers have a long life and since many DXers trade up to a higher-priced set than the one they have, a large number of used receivers are on the market. Many dealers take in older but perfectly good radios in trade. Many will recondition the set before reselling it. Others will simply "sell as is." If it needs work, but it is basically in good condition, you can recondition it yourself or have it done. Dealing with a reputable dealer will minimize the risks of getting a "lemon."

Most American manufacturers will realign and recondition an older set they have manufactured and the result will be more than satisfactory. This service costs a modest amount, but if you bought the unit at the right price, it could be well worth it.

Depending on the condition of the unit and the demand for a particular brand, you can probably save a quarter or half of the original price when buying "second hand."

Once you have chosen and purchased a set, you ask the question, "How do I set it up?" This is not too difficult to do.

## SETTING UP YOUR DX STATION

Now that you have gathered your accessories and operating aids, the big question is: Where do we put all this equipment and in what way should we do that?

A corner arrangement as shown in Fig. 2-5 allows for easy and efficient access to the equipment and allows for expansion into amateur radio, if desired. Of course, you can set up your receiving station in many different places around the house: up in the attic, in a spare bedroom, in a corner of the living room, in a shed in the yard, or in a cellar. Whatever location you choose, it will depend on your personal circumstances and the opinion of others in your household. Where you are going to install it is not as important as the arrangement itself, since your DX hobby should remain a pleasure to execute. So, heating and lighting are important. A quiet place

Fig. 2-5. A simple station arranged in a corner.

where you will not interfere with the activities of others in the house is ideal. Wherever that will be, you will have to take some measurements for practical and safety reasons. You should arrange for a power connection that has its own safety fuse to prevent the whole house from being darkened in case of a short or overload. You should also look for sufficient power connections to take care of all your receivers and accessories. A "cube-tap octopus arrangement" like that shown in Fig. 2-6 might cause a dangerous circuit overload and should be avoided.

Also install an antenna panel in such a way that you can easily remove the plugs when you leave the installation. This will prevent possible damage to your equipment in case thunderstorms occur during your absence.

To prevent tampering by small children and unauthorized persons, you should also install a central power switch, located as high as possible. Otherwise, just use a normal table and place your equipment on it. Leave some room for your logbook and an ashtray, if you are a smoker. Furthermore there is no harm in placing certain pieces of equipment on top of each other. When you begin expanding and you have gathered too much material, you might consider installing shelves against the wall behind the table to hold your accessories and additional receivers.

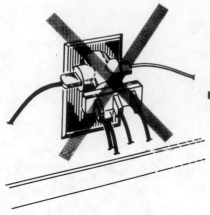

Fig. 2-6. Connecting more than two appliances into one outlet is extremely dangerous.

Remember one thing, however—whatever setup you choose, do it according to your own taste and for your own pleasure.

## OPERATING A DX STATION

There's an art to tuning a receiver. Do you know what to do with all the controls on a set? Do you know what they mean? Simply roaming at random across the dial may be easy, but it will not produce satisfactory results, especially if you want to hear distant stations.

The first step, therefore, is to learn the purpose and function of all the knobs on the receiver. Each manufacturer has the purpose and function of the controls clearly covered in the instruction manual. Read the manual carefully. Practice with all the controls until you understand them fully. Then begin listening *seriously* on the lowest frequency shortwave band. The signals are stronger on this band and less prone to fading than those in the higher frequencies.

Tune the main frequency dial slowly and listen to how the stations come in to audibility and fade out again. Note how some stations are spaced closely together at some parts of the dial and farther apart at others. You will notice that a shortwave station occupies a smaller fraction of the dial than a regular am broadcast station. You, therefore, have to tune very carefully and very slowly, because you can skip over a shortwave station when tuning too fast.

Practically all shortwave receivers incorporate *automatic volume control* to reduce the effects of fading and auroral flutter. Use this aid especially when listening to broadcast signals. Sometimes, you can also reduce interference by manipulating the AUDIO TONE CONTROL. High-pitched signals between two adjacent stations can be reduced by turning the control more towards bass reception. To allow good reproduction of speech you should turn this control more towards treble, however.

Next, turn the BAND SWITCH to a high frequency and practice tuning foreign broadcast stations. Doing so, you will notice that it is much harder to tune in a weak distant signal than it is a strong local one. Little-by-little you will gain skill with your receiver.

## OPERATING A DX RECEIVER

There is one control that requires some practice before you can totally master it—the BANDSPREAD TUNER.

Bandspread refers to the ability of a receiver to separate stations on a dial without having an accumulation of signals at a certain point.

Most modern receivers employ bandspread tuning whereby a small portion of a frequency spectrum may be spread over a large segment of the dial for ease of tuning and logging accuracy. The bandspread may be mechanical, using a gear system that permits a slow rate of tuning with an auxiliary tuning knob, or it may be electrical.

Using the receiver shown in Fig. 2-7 as a reference, bandspread operation is accomplished as follows:

(1) Turn BAND switch to band E.
(2) Using MAIN TUNING knob, set the dial at 14 MHz.

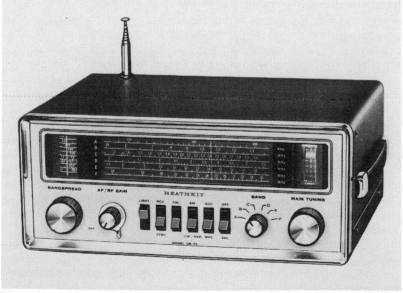

Courtesy Heath Co.

**Fig. 2-7. Model GR-78 transistorized general-coverage receiver.**

(3) Look at the left side of the set. You will notice a control labeled BANDSPREAD. Above it is another tuning element with a number of frequencies listed. Look closely, and you will notice on the third scale from the bottom a set of numbers, starting at 14.0. This scale—actually calibrated to a set of tuning capacitors—allows you to spread the 14-MHz band into separate frequencies.

Any time you move the MAIN TUNING control, however slightly, you can spread the frequency with the BANDSPREAD control.

# CHAPTER 3

# Accessories and Operating Aids

In order to operate your DX station in a more convenient manner and to enhance the pleasure of listening, you need certain accessories. This chapter will cover several accessories including: headphones, recorders, tuners, filters, maps, and logbooks. There are, of course, numerous other accessories available and as you gain experience and the need you will discover them yourself.

## HEADPHONES

This is perhaps the first additional item you should acquire. Its utility borders on necessity in some instances. A headphone set will obviously be appreciated by those members of your household who may not appreciate the squalls and squawks of shortwave radio. A more scientific reason for headphones is the advantage of some filtering that they provide. Communications headphones are vastly more desirable than hi-fi headphones. The reason for this is so that you can restrict the frequency response of your DX to tones approximating those of the human voice—the lower tones of buzzing noises and the higher noises of the squalls can be attenuated. High-fidelity headphones pass on the wide range and fidelity of music and voice— as well as the fidelity of crackles, pops, and buzzes.

Carbon-type headphones are the most ideal. Also, magnetic types are acceptable. The moving-coil headphones, however, should be avoided. Headphones have different ratings, such as 8 ohms, 600 ohms, or 2,000 ohms. Determine what impedance—in ohms—your receiver offers at its headphone jack—examine your manual for specification—and purchase accordingly. If you are unable to make this determination, then 2,000-ohm headphones may be used. Damage to your headphones would probably only occur when using 8-ohm headphones in a high-impedance phone outlet of 2,000 ohms at high volume. The fine gauge wires of your headphones or the audio transformer might not be able to withstand the heavy current.

## 24-HOUR CLOCK

Since DXing deals mainly with GMT, or UT (Universal Time), which is a 24-hour system, you should obtain a clock that will tell time at a glance.

*Note: Some DXers are using their local time for keeping their logs and this gets on the QSL cards. Picture yourself in Outer-Mongolia trying to equate Eastern U.S. daylight time with GMT so you can locate the contact in your log?!! So, keep your clock and log in GMT.*

## TAPE RECORDER

This is another semi-essential tool for the DXer gaining experience. Sooner or later, recorded reports and tapes for your library become desirable. Although cassette tape recorders are very attractive and easy to use, it is recommended that any recorded reception reports be sent on a small reel of tape. Actually, if you can afford both a reel and a cassette tape recorder, you will be at a great advantage.

There are many and varied kinds of recorders, but since shortwave is monaural sound, it will be a wasted expense to purchase stereo recording equipment. A simple reel-to-reel recorder—even one that accepts only 3-inch tapes—will suffice nicely. Make sure that your recorder is a capstan-driven machine capable of playing either at $3\frac{3}{4}$ i.p.s. or $1\frac{7}{8}$ i.p.s. and that it has a digital tape counter.

Other features your recorder should have include: the capability to operate from either battery or regular current; an external earphone plug—to provide you with the ability to listen through while recording a reception; and an auxiliary plug-in socket to allow recording from a speaker source.

Most modern receivers are equipped with a RECORDER socket to which you can connect a tape recorder. This direct hookup provides you with the best quality reproduction of the recording. If your receiver does not have a RECORDER socket, then you could connect your tape recorder to the speaker outlets of your receiver.

Although we will deal with reception reporting in another chapter, a few words of caution regarding the mailing of recorded tapes. When tape reporting, you should always make out your normal letter report, except where details of the programming are to be mentioned, then you need only to refer to the enclosed tape recording. Inquire by letter first, whether the broadcasting station will accept tapes for identification. Always send reel-to-reel tapes, recorded at 3¾ i.p.s. on ONE side only. The reverse side must be blank.

Although X-rays are used by the postal department to inspect parcels, etc., this does not erase the recordings on your tape. It may be well nevertheless to indicate on the package that tapes are enclosed.

One final note on recording is needed. Often hobbyists attempt to record programs, etc., at too loud a volume. A signal that is too strong may cause the metallic-oxide surfaces of the tape to become over magnetized. This, in turn, will create distortions and other unwanted signals during playback or when re-recording the tape.

## ANTENNA TUNERS

The antenna tuner is a device designed to make the antenna resonate with optimum results. Some tuners are designed to amplify a signal before it is sent to the receiver—such tuners are referred to as preamplifiers or preselectors. There is some difference between the two types. The tuner uses a very broad tuning and amplifying circuit that magnifies and amplifies all signals, including noise, in a given small band of frequencies. A preselector/tuner has the ability to discriminate between signals and man-made noise and is more critical in its tuning.

## NOISE FILTERS

A noise filter, connected externally to your receiver, is a circuit used to suppress man-made noises generated from car ignitions, power lines, electrical appliances, etc.

## AUDIO FILTERS

An audio filter is installed between the receiver and the headphones. Its function is to restrict frequency responses in the narrow band centering on 800 Hz—cw. It filters out frequencies lower and higher than the center frequency, allowing the center frequency to come through louder and clearer.

## MAPS

Maps are a great aid to you as a DXer. They will help you to learn more about the countries and peoples whose signals you are receiving on your set. The National Geographic Society has over 50 different maps available that chart just about any part of the world in detail. You can get a list of these from the National Geographic Society, Washington, DC 20036.

The Radio Amateur Callbook publishes a Radio Amateurs World Atlas that has separate maps of Europe, Asia, Africa, North America, South America, the Pacific Area, and a polar projection of the world. They also have maps printed in four colors available, showing the call prefixes and zones of the world.

You could also invest in a better-type atlas that is available from $25 up. Such an atlas not only has detailed maps of every area of the world, but also has a gazetteer that enables you to look up any town in the world and find out where it is and what population it has. When a broadcaster in the wilds of Siberia spells out the name of his town you will be able to find out exactly where it is.

Having a globe right beside your station setup is a very handy operating aid for quick figuring of beam angles and just to look at when you listen to someone from an exotic place on the other side of the world.

## LOGBOOKS

Since his beginning, man has recorded events for posterity in one form or another. Records have been kept of virtually everything that has ever happened and some of these records are still in existence today, kept in hermetically sealed museum display cases and time capsules. You can still check with the state and national government archives to learn about something that happened years ago.

It all means that it is worthwhile, and often very necessary, to keep a record—a log—of what was or is happening. So it is with DXing. A logbook provides you with answers to questions that come up from time to time about the stations you have heard in the past.

It is not really necessary to keep a log of every last thing that you hear. In other words, if you are just casually tuning across one of the bands and you happen to stop momentarily on a half dozen or so stations and spend only a moment or two for a brief glimpse of the program, it is not necessary to record all of that information. The general rule of thumb is that the logbook should contain material about any station to which you might want to send a reception report. You may also want to record information—even though it may be for a brief logging—about stations that you do not normally hear, such as stations that only come through during ionospheric disturbances.

A logbook is also a good source of information in helping you plan your listening time. For example, if you want to try to hear a certain station in Africa, you can go back in your logbook for the past few weeks and determine what time that area was last coming through. Chances are that unless a long period of time has elapsed since your last logging or unless there were some propagation conditions, you will again hear the same general area at about the same time of day or night.

There are several kinds of logs; each has its distinct advantages and disadvantages.

One method of logging is to list stations according to their frequencies. A page is set aside for a certain band of frequencies—including say 11,750–11,800 kHz—and all stations heard on these frequencies are listed on that page (see Fig. 3-1). The frequency span

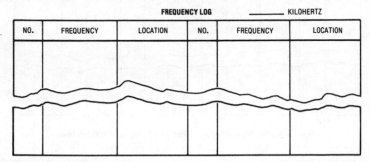

Fig. 3-1. A typical page for keeping a frequency log of stations received.

is listed in the upper right hand corner and the entire collection of pages is kept in order of frequency. You may use dividers to separate frequency bands. The "No." column is used to keep track of the number of stations logged on a particular frequency. Start with the number 1 and continue your listing. The advantage of the frequency log is the easy access you will have to information about the stations you heard on a particular frequency.

Another logbook is the "Station Log"—"Ascension Log"—in which you record the stations in the order in which you hear them (see Fig. 3-2). It is also used as a basic reference log. Since the station log is an important document to you as a DXer, it contains a great deal of information. The frequency, call, location, time, and date are the basic entries made in this logbook. Time is, of course, entered using the 24-hour system. The column marked "signal" could be given a SINPO rating. See Chapter 6 for an explanation of various signal strength and quality reporting codes. The "QRM" might list both the interfering station and if the interference was sideband splatter, also the frequency of that station. The abbreviation "RPT" is for *reported* and only a checkmark ($\checkmark$) needs to be entered in this column, if a QSL was requested. The column headed "QSL'D" is short for *QSL received*. Column "TP'D" means *taped* and a simple checkmark ($\checkmark$) or some form of coding to indicate the reel number can be entered in this column. The "Remarks" column is too small to be used to enter program notes. Instead you may want to use it for entering such notes as *new frequency* or the like.

The third log, the "Report Log" shown in Fig. 3-3, supports the Station Log. It can be used to enter specific details about certain

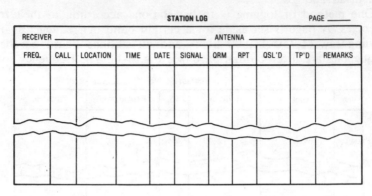

Fig. 3-2. A typical page for keeping a station log.

receptions. On the "Interference" line you should state the interfering station or atmospheric interference. The "Readability" line might contain a percentage figure to indicate how much of what was received was intelligible.

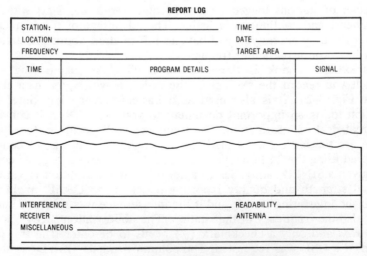

Fig. 3-3. A typical page for a report log.

There are surely numerous other ways of record keeping. However, the sample logs shown here are in essence the logs used by experienced DXers and they have proven to be indispensable.

## STATION LOCATION AND PREFIX GUIDE

| Country/Area | Continent | Zone | Prefix | Station Site |
|---|---|---|---|---|
| Aden | A | 21 | VS9 | Aden |
| Afghanistan | A | 21 | YA | Kabul |
| Aland Islands | E | 15 | OH0 | |
| Alaska | NA | 1 | KL7 | Anchorage |
| Albania | E | 15 | ZA | Tirana |
| Algeria | AF | 33 | 7X | Algiers |
| Andorra | E | 14 | PX1 | Andorre-la-Vielle |
| Angola | AF | 36 | CR6 | Luanda |
| Antigua | NA | 8 | VP2A | St. Johns |
| Argentina | SA | 13 | LU | Buenos Aires, etc. |
| Ascension Islands | AF | 36 | ZD8 | Ascension |
| Australia | O | 29/30 | VK | Perth, Sidney, etc. |
| Austria | E | 15 | OE | Vienna, etc. |
| Azores Islands | E | 14 | CT2 | Ponta Delgada |
| Bahamas | NA | 8 | VP7 | Nassau |
| Balearic Islands | E | 14 | EA6 | Majorca, etc. |
| Barbados Islands | NA | 8 | 8P6 | Bridgetown |
| Belgium | E | 14 | ON | Brussels |
| Bhutan | A | 22 | AC5 | Thimpu |
| Bolivia | SA | 10 | CP | LaPaz, etc. |
| Borneo | O | 28 | 8F5 | Pontianak |
| Brazil | SA | 11 | PY | Rio de Janeiro, etc. |
| British Honduras (Belize) | NA | 7 | VP1 | Belize |
| Brunei | O | 28 | VS5 | Berakas, etc. |
| Bulgaria | E | 20 | LZ | Sofia |
| Burma | A | 26 | XZ | Rangoon |
| Burundi | AF | 36 | 9U5 | Bujumbura |
| Cambodia | A | 26 | XU | Phnom Penh |
| Cameroun | AF | 36 | TJ | Yaounde |
| Canada | NA | 1–5 | VE,3C | Montreal |
| Canal Zone | NA | 7 | KZ5 | Balboa |
| Canary Islands | AF | 33 | EA8 | S. Cruz de Teriffe |
| Cape Verde Islands | AF | 35 | CR4 | Sao Vicente, etc. |
| Celebes/Molucca | O | 28 | 8F6 | Ujung Pandang |
| Chad | AF | 36 | TT8 | Ndjamana |
| Channel Islands | E | 14 | GC | Jersey |
| Chile | SA | 12 | CE | Santiago, etc. |
| China (mainland) | A | 23/24 | BY | Peking, etc. |
| Colombia | SA | 9 | HK | Bogota, etc. |
| Comoro Islands | AF | 39 | FH8 | Moroni, etc. |

| Country/Area | Continent | Zone | Prefix | Station Site |
|---|---|---|---|---|
| Cook Islands | O | 32 | ZK1 | Raratonga, etc. |
| Costa Rica | NA | 7 | TI | San Jose, etc. |
| Cuba | NA | 8 | CM/CO | Havana, etc. |
| Cyprus | A | 20 | 5B4/ZC4 | Limassol, etc. |
| Czechoslovakia | E | 15 | OK | Prague |
| Dahomey | AF | 35 | TY | Cotonue |
| Denmark | E | 14 | OZ | Copenhagen |
| Dodecanese Islands | E | 20 | SV5 | Rhodes |
| Dominican Rep. | NA | 8 | HI | Santo Domingo, etc. |
| Equador | SA | 10 | HC | Quito, etc. |
| Egypt | AF | 34 | SU | Cairo |
| El Salvador | NA | 7 | YS | San Salvador |
| England | E | 14 | G | London, etc. |
| Ethiopia | AF | 37 | ET3 | Addis Ababa |
| Falkland Islands | SA | 13 | VP8 | Port Stanley |
| Fiji Islands | O | 32 | VR2 | Suva |
| Finland | E | 15 | OH | Helsinki |
| France | E | 14 | F | Paris, etc. |
| Gabon | AF | 36 | TR8 | Libreville |
| French Guinea | SA | 9 | FY7 | Cayenne |
| Galapagos Islands | SA | 10 | HC8 | Pto. Baq. Moreno |
| Gambia | AF | 35 | ZD3 | Benjul |
| Germany, West | E | 14 | DJ/DK/DL | Cologne, etc. |
| Germany, East | E | 14 | DM | East Berlin |
| Ghana | AF | 35 | 9G1 | Accra |
| Gilbert Islands | O | 31 | VR1 | Tarawa |
| Greece | E | 20 | SV | Athens, etc. |
| Greenland | NA | 40 | OX | Odthaab |
| Grenada | NA | 8 | VP2G | St. George's |
| Guadeloupe Islands | NA | 8 | FG7 | Pointe-a-Pitre |
| Guatemala | NA | 7 | TG | Guatemala City |
| Guiana | SA | 9 | FY7 | Conakry |
| Haiti | NA | 8 | HH | Port-au-Prince |
| Hawaii | O | 31 | KH6 | Honolulu |
| Honduras | NA | 7 | HR | Tegucigalpa |
| Hong Kong | A | 24 | VS6 | Victoria |
| Hungary | E | 15 | HA/HG | Budapest |
| Iceland | E | 40 | TF | Reykjavik |
| India | A | 22 | VU | New Delhi |
| Indonesia | O | 28 | 8F | Jacarta, etc. |
| Iran | A | 21 | EP/EQ | Teheran, etc. |
| Iraq | A | 21 | YI | Baghdad |
| Ireland | E | 14 | EI | Dublin |
| Israel | A | 20 | 4X4/4Z4 | Jerusalem |
| Italy | E | 15 | I | Rome, etc. |
| Ivory Coast | AF | 35 | TU2 | Abidjan |
| Jamaica | NA | 8 | 6Y5 | Kingston |
| Japan | A | 25 | JA/JH | Tokyo, etc. |

54

| Country/Area | Continent | Zone | Prefix | Station Site |
|---|---|---|---|---|
| Jordan | A | 20 | JY | Amman |
| Kenya | AF | 37 | 5Z4 | Nairobi |
| Korea | A | 25 | HM/HL9 | Seoul |
| Kuwait | A | 21 | 9K2 | Kuwait |
| Laos | A | 26 | XW8 | Vientiane |
| Lebanon | A | 20 | OD5 | Beirut |
| Liberia | AF | 35 | EL | Monrovia |
| Libya | AF | 34 | 5A | Tripolis |
| Liechtenstein | E | 14 | HB0 | |
| Luxembourg | E | 14 | LX | Villa Louvigny |
| Macao | A | 24 | CR9 | |
| Malagasy | AF | 39 | 5R8 | Tananarive |
| Malaysia | A | 28 | 9M2 | Kuala Lumpur |
| Maldive Islands | A | 22 | VS9M | Male |
| Mali Republic | AF | 35 | TZ | Bamako |
| Malta | E | 15 | 9H1 | Valetta |
| Manchuria | A | 24 | BY9 | Harbin |
| Martinique | NA | 8 | FM7 | Fort-de-France |
| Mauritania | AF | 35 | 5T5 | Nouakchott |
| Mauritius Islands | AF | 39 | VQ8 | Forest Side |
| Mexico | NA | 6 | XE/XF | Mexico City |
| Monaco | E | 14 | 3A2 | Monte Carlo |
| Mongolia | A | 23 | JT1 | Ulan Bator |
| Morocco | AF | 33 | CN | Rabat |
| Mozambique | AF | 37 | CR7 | Lorenco Marques |
| Nepal | A | 22 | 9N1 | Katmandu |
| Netherlands | E | 14 | PA/PI | Hilversum |
| Netherlands Antilles | SA | 9 | PJ | Bonaire |
| New Caledonia | O | 32 | FK8 | Noumea |
| Newfoundland | NA | 5 | VO1/3B1 | St. Johns |
| New Hebrides | O | 32 | YJ | Vila |
| New Zealand | O | 32 | ZL | Wellington |
| Nicaragua | NA | 7 | YN | Managua |
| Nigeria | AF | 35 | 5N2 | Lagos |
| Norway | E | 14 | LA | Oslo |
| Pakistan | A | 21/22 | AP | Karachi |
| Panama | NA | 7 | HP | Panama City |
| Papua | O | 28 | VK9 | Raboul |
| Paraguay | SA | 11 | ZP | Ascuncion |
| Peru | SA | 10 | OA | Lima |
| Philippines | O | 27 | DU | Manilla |
| Poland | E | 15 | SP | Warsaw |
| Portugal | E | 14 | CT1 | Lisbon |
| Qatar | A | 21 | MP4Q | Doha |
| Reunion Islands | AF | 39 | FR7 | St. Denis |
| Rumania | E | 20 | YO | Bucharest |
| Rwanda | AF | 36 | 9X5 | Kigali |
| Saudi Arabia | A | 21 | HZ/7Z | Jeddah |

| Country/Area | Continent | Zone | Prefix | Station Site |
|---|---|---|---|---|
| Scotland | E | 14 | GM | Edinburgh |
| Seychelles Islands | AF | 39 | VQ9 | Victoria |
| Sicily | E | 15 | IT | Caltanisetta |
| Sierra Leone | AF | 35 | 9L1 | Freetown |
| Singapore | A | 28 | 9V1 | Singapore |
| Solomon Islands | O | 28 | VR4 | Honiara |
| Somali Rep. | AF | 37 | 6O | Mogadishu |
| South Africa | AF | 38 | ZS | Paradys, etc. |
| Spain | E | 14 | EA | Madrid, etc. |
| Spanish Sahara | A | 33 | EΛ9 | Aguin |
| Surinam | SA | 9 | PZ | Paramaribo |
| Swaziland | AF | 38 | ZD5 | Mbane |
| Sweden | E | 14 | SL/SM | Stockholm |
| Switzerland | E | 14 | HB | Berne |
| Syria | A | 20 | YK | Damascus |
| Taiwan | A | 24 | BV | |
| Tanzania | AF | 37 | 5H3 | Dar-es-Salaam |
| Thailand | A | 26 | HS | Bangkok |
| Tibet | A | 23 | AC4 | Lhasa |
| Togo | AF | 35 | 5V | Lome |
| Trinidad | SA | 9 | 9Y4 | Port of Spain |
| Tunesia | AF | 33 | 3V8 | Tunis |
| Turkey | A | 20 | TA/TC | Ankara |
| Uganda | AF | 37 | 5X5 | Kampala |
| USA | NA | 3/4/5 | W/K | Greenville, etc. |
| USSR | E | 15 | UA1, 3, 4, 6 | Moskou, etc. |
| Upper Volta | AF | 35 | XT2 | Uoagadougou |
| Uruguay | SA | 13 | CX | Montevideo |
| Vatican City | E | 15 | HV | Vatican |
| Venezuela | SA | 9 | YV | Caracas |
| Vietnam | A | 26 | 3W8 | Hanoi, Saigon |
| Yemen | A | 21 | 4W1 | San'a |
| Yugoslavia | E | 15 | YT/YU | Belgrade |
| Zambia | AF | 36 | 9J | Lusaka |

The map in Fig. 3-4 shows the world divided into 40 geographic zones for reporting stations received from various areas.

## TIME DIFFERENCE AROUND THE WORLD

To compute the time anywhere in the world, add (+) or subtract (−) the number of hours indicated to or from Greenwich Mean Time (GMT). The figures in parentheses indicate Daylight Savings Time used during the summer months in a particular country.

| Country/Area | Hours from GMT | Country/Area | Hours from GMT |
|---|---|---|---|
| Aden | +3 | Albania | +1 |
| Afghanistan | +4½ | Algeria | +1 |

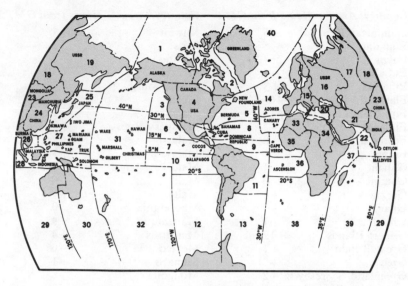

Fig. 3-4. World map showing 40 geographic location zones.

| Country/Area | Hours from GMT | Country/Area | Hours from GMT |
|---|---|---|---|
| Angola | +1 | Eastern | −5(−4) |
| Argentina | −4(−3) | Central | −6(−5) |
| Ascension Islands | 0 | Mountain | −7(−6) |
| Australia–Eastern | +10 | Pacific | −8(−7) |
| Central | +9½ | Yukon | −9(−8) |
| Western | +8 | Canary Islands | 0 |
| Austria | +1 | Cape Verde Islands | −2 |
| Azores | −1 | Central African Rep. | +1 |
| Bahamas | −5 | Chad | +1 |
| Bahrein | +4 | Chile | −4 |
| Belgium | +1 | China–Tihwa–Zone 1 | +6 |
| Bermuda | −4 | Chungking | +7 |
| Bolivia | −4 | Peking/ | |
| Botswana | +2 | Shanghai | +8 |
| Brazil–Eastern | −3(−2) | Harbin–Zone 4 | +8½ |
| Central | −4(−3) | Cocs Islands | +6½ |
| Western | −5(−4) | Colombia | −5 |
| Brunei | +8 | Cook Islands | −10½ |
| Bulgaria | +2 | Corsica | +1 |
| Burma | +6½ | Costa Rica | −6 |
| Burundi | +2 | Cuba | −5 |
| Cambodia | +7 | Cyprus | +2 |
| Cameroun | +1 | Czechoslovakia | +1 |
| Canada–Newfoundland | −3½ | Dahomey | +1 |
| Atlantic | −4(−3) | Denmark | +1 |

| Country/Area | Hours from GMT | Country/Area | Hours from GMT |
|---|---|---|---|
| Dominican Rep. | −5 | Kuwait | +3 |
| Equador | −5 | Laos | +7 |
| Egypt | +2(+3) | Lebanon | +2 |
| El Salvador | −6 | Liberia | +3½ |
| Ethiopia | +3 | Luxembourg | +1 |
| Flakland Islands | −4(−3) | Libya | +2 |
| Fiji Islands | +12 | Madeira | −1 |
| Finland | +2 | Malagasy Rep. | +3 |
| Formosa | +8 | Malawi | +2 |
| France | +1 | Malaysia | +7½ |
| Gabon | +1 | Maldive Islands | +5½ |
| Gambia | 0 | Mali | 0 |
| Germany | +1 | Malta | +1 |
| Ghana | 0 | Mauretania | 0 |
| Gibraltar | +1 | Martinique | −4 |
| Gilbert Islands | +12 | Mauritius | +4 |
| Great Britain | +1 | Mexico | −6 |
| Greece | +2 | Monaco | +1 |
| Greenland–Scoresby S. | −2 | Mongolia | +8 |
| West Coast | −3 | Morocco | 0 |
| Thule | −4 | Mozambique | +2 |
| Guam | +10 | Nepal | +5½ |
| Guatemala | −6 | Neth. Antilles | −4 |
| Guiana, Fr. | −3 | New Caledonia | +11 |
| Guinea | 0 | New Guinea | +10 |
| Guinea, Equator. | +1 | New Hebrides | +11 |
| Gyana | −3½ | New Zealand | +12 |
| Haiti | −5 | Nicaragua | −6 |
| Holland | +1 | Niger | +1 |
| Honduras | −5(−6) | Nigeria | +1 |
| Hong Kong | +8(+9) | Norfolk Islands | +11½ |
| Hungary | +1 | Norway | +1 |
| Iceland | −10 | Nyasaland | +2 |
| India | +5½ | Pakistan | +5 |
| Indonesia–West | +7 | Panama | −5 |
| Central | +8 | Pappua–New Guinea | +10 |
| East | +9 | Paraguay | −4 |
| Iran | +3½ | Peru | −5 |
| Iraq | +3 | Philippines | +8 |
| Ireland | 0 | Poland | +1 |
| Israel | +2 | Portugal | +1 |
| Italy | +1(+2) | Puerto Rico | −4 |
| Ivory Coast | 0 | Qatar | +4 |
| Japan | +9 | Reunion Islands | +4 |
| Jordan | +2 | Rhodesia | +2 |
| Kamaran Islands | +3 | Rumania | +2 |
| Kenya | +3 | Rwanda | +2 |
| Korea | +9 | Ryuku Islands | +9 |

| Country/Area | Hours from GMT | Country/Area | Hours from GMT |
|---|---|---|---|
| Sao Tome | 0 | Sverdlovsk | +5 |
| Samoa | −11 | Tashkent | +6 |
| Saudi Arabia | +3 | Novosibirisk | +7 |
| Sarawak | +8 | Irkutsk | +8 |
| Senegal | 0 | Yakutsk | +9 |
| Seychelles | +4 | Khabarovsk | +10 |
| Sierra Leone | 0 | Magadan | +1 |
| Singapore | +7½ | Petropavlosvk | +12 |
| Solomon Islands | +11 | Anadyr | +13 |
| Somali Rep. | +3 | USA–Eastern | −5(−4) |
| South Africa | +2 | Central | −6(−5) |
| Spain | +1 | Mountain | −7(−6) |
| Sri Lanka | +5½ | Pacific | −8(−7) |
| Sudan | +2 | Yukon | −9(−8) |
| Surinam | −3¼ | Alaska–Juneau | −8 |
| Sweden | +1 | Anchor- | |
| Switzerland | +1 | age | −10 |
| Syria | +2(+3) | Aleu- | |
| Tahiti | −10 | tians | −11 |
| Tanzania | +3 | Hawaii | −10 |
| Thailand | +7 | Bering | −11 |
| Timor | +8 | Vatican | +1 |
| Togo | 0 | Venezuela | −4 |
| Tonga | +1 | Vietnam | +7 |
| Tunisia | +1 | Windward Islands | −4 |
| Turkey | +2 | Yemen | +1 |
| Uganda | +3 | Yugoslavia | +1 |
| Upper Volta | 0 | Zaire–West | +1 |
| Uruguay | −3 | East | +2 |
| USSR–Moscow | +3 | Zambia | +2 |
| Baku | +4 | | |

# CHAPTER 4

# Antennas

The antenna has always been an important part of the DXer's equipment, as many have found out just how helpful a good antenna can be. The majority of DXers, however, still employ a relatively simple antenna: an outdoor wire of a certain length, a rod antenna, or the built-in telescoping antenna of a portable receiver.

The job of your antenna is to capture electromagnetic energy transmitted from a distant station and convert it into an electron flow to your receiver. Then, in your receiver it is changed to an audible signal. A good antenna can pull in weak stations, as well as give you more reliable reception of stronger stations. Therefore, your antenna should receive just as much attention, if not more, as the selection of your receiver. Whether you own a $20.00 bargain special or a high-priced set, reception will be greatly improved by adding an outdoor antenna to your equipment.

What type of antenna should you use? How long should it be? How high? Indoor or outdoor? What about directionality? Antenna design is a complex subject. There are a number of books available to help you with detailed information on constructing antennas. A receiving-only antenna, however, is a lot less critical than one that is used for transmitting. Your basic question, of course, is, "What would be the ideal DXing antenna?" Well, it would be a broad-band antenna, that is, it would work well on all or, at least, most of the shortwave frequencies—from the bottom to the top of the dial.

So, a good antenna *is* the key to successful shortwave reception. Time and effort spent in making the best antenna installation possible will result in greatly improved reception. No matter what form or type of antenna you use, it must follow some fundamental rules.

(1) The antenna should be erected as high as possible. Twenty feet above the ground may be considered to be a minimum height.

(2) The location of the antenna with respect to nearby objects is very important. Telephone lines and power lines exhibit a shielding effect to a nearby antenna. Stucco buildings with wire mesh in the walls are also a detriment to best reception. If possible, place your antenna so that its length is at right angles to nearby utility wires. Keep your antenna clear of the walls of nearby buildings, or try to place the antenna above the top of the building.

(3) The antenna installation should be sturdy and well made. Occasionally you will read of a radio antenna falling across a high-tension wire and electrocuting the installer. You should avoid make-shift connections when you erect an antenna. Stay away from power lines and make sure that if your antenna should come down, it will not hit a utility cable! If the antenna is attached to a tree as shown in Fig. 4-1, some allowance should be made for the sway of the tree in the wind. A rope, pulley, and weight should be used.

The numbers in the following construction details correspond to the numbered items called out in Fig. 4-1.

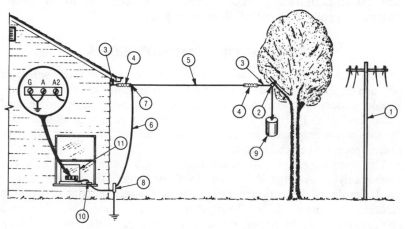

**Fig. 4-1. Long-wire antenna is easy to construct.**

Place your antenna at a right angle to any nearby telephone and utility power lines (1).

Attach a pulley (2) to a tree branch with a rope (3).

Fasten the rope to one end of an insulator (4).

Connect your antenna wire—No. 14 enameled copper wire, 20 to 100 feet long (5)—to the insulator and attach an insulator (4) to the other end.

Wire twist and solder both ends of your antenna wire.

Connect lead-in wire (6)—No. 14 insulated copper wire, if possible—to one end of antenna (7).

Fasten other end of lead-in wire to lightning arrestor (8).

Ground the lightning arrestor (8).

If your lead-in wire (6) is not insulated, keep it away from the building and window sill by using a lead-in strip (10) when guiding the wire through the window.

Connect grounded side of the arrestor to A2 of your receiver (11) antenna connection, as shown.

The counter weight (9) will help keep your antenna straight when the wind blows.

Your inverted "L" or "long-wire antenna" is now ready to provide radio signals.

All joints in the antenna wires should be well soldered. In windy locations where the antenna could come down, steel-core copper wire should be used with strain insulators at each end.

(4) In locations having a great deal of moisture or industrial smoke, enameled copper wire should be used for antenna construction. As an additional protective measure, all soldered joints should be given a protective coat of aluminum paint.

## KINDS OF TRANSMISSION LINES

Since the length and material used for your antenna transmission line are important considerations when constructing your system, you should understand something about some of the kinds of wire and cable available. Fig. 4-2 shows three different communications conductors commonly used in receiving antenna systems: open wire, tv ribbon wire, and coaxial cable.

The open-wire line with insulating spacers would be suitable for your receiving antenna if No. 14 AWG or larger size copper wire is used. Insulating spacers are available in various lengths to provide suitable separation; 8-inch insulators should be adequate. The two conductor tv lead-in ribbon or twin lead is a polyethylene insulating plastic around two copper wire leads. This conductor commonly has an impedance value of 300 ohms and is suitable for carrying signals in the tv broadcast bands. Such wire is not ordinarily used for the antenna but, rather, for bringing the signal from the antenna to your receiver. Fig. 4-2C shows another kind of lead-in cable which is widely used and highly recommended. It is called coaxial cable and is strong and also flexible, allowing it to bend around corners or over window sills, etc. Coaxial cable has a solid copper center conductor with a polyethylene dielectric insulation covered by a braided metal

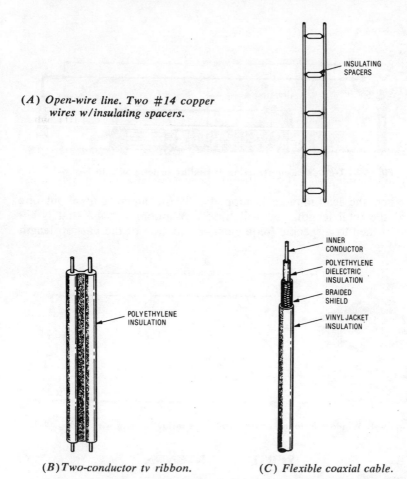

(A) Open-wire line. Two #14 copper wires w/insulating spacers.

INSULATING SPACERS

INNER CONDUCTOR

POLYETHYLENE DIELECTRIC INSULATION

BRAIDED SHIELD

VINYL JACKET INSULATION

POLYETHYLENE INSULATION

(B) Two-conductor tv ribbon.          (C) Flexible coaxial cable.

**Fig. 4-2. Example of antenna and transmission lead-in lines.**

shield and a vinyl outer jacket. The braided shield, when properly grounded, carries off noise and unwanted spurious signals and leaves the signal received on the inside conductor "clean" to be carried to your receiver. Coaxial cable of this kind commonly is impedance rated at 52 ohms.

## T ANTENNA AND WINDOM ANTENNA

You may want to connect the lead-in wire near the center of the antenna rather than at the end (see Fig. 4-3). This is called a T antenna. The horizontal wire—or flat top—should be about 90 feet long and the lead-in wire can be any reasonable length.

63

Fig. 4-3. T-antenna construction is similar to long wire in Fig. 4-1.

When the lead-in wire is tapped into the antenna at about one third the total length, you will have a Windom-antenna that is actually tuned to a specific frequency—a function of the flat-top length (see Fig. 4-4).

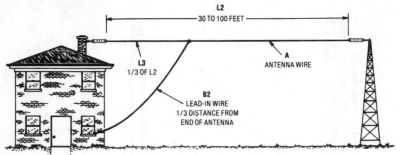

Fig. 4-4. Windom antenna construction is similar to long wire in Fig. 4-1.

## VERTICAL ANTENNA

The vertical antenna is relatively nondirectional—it receives equally in all directions. This antenna configuration is very responsive to atmospheric and man-made noise—much more so than the horizontal antenna. However, compared with the horizontal antenna, it takes up very little space (see Fig. 4-5). A vertical antenna can be attached to a chimney, or to the side of a house or tree. If you live in an area that has severe automobile ignition noise and your static level is reasonably low, you might think about using a vertical antenna which works well over the entire frequency spectrum.

## DIPOLE ANTENNA

The internationally adopted standard antenna is the half-wave dipole. It is a *resonant* antenna tuned to a narrow band of frequen-

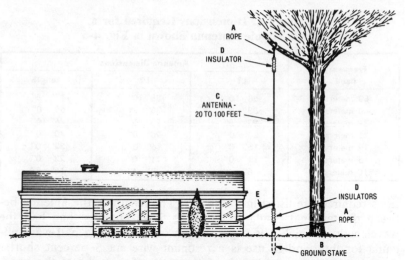

**Fig. 4-5. Vertical antenna construction is similar to long wire in Fig. 4-1.**

cies. For any antenna there is one frequency, called the *resonant frequency,* at which point the inductive and capacitive reactance properties of the system become exactly equal and neutralize each other, allowing maximum amplitude of the frequency to be passed to the receiver. Antenna current, therefore, is largest at the resonant frequency.

Basically the dipole antenna shown in Fig. 4-6 consists of two wires of equal length, separated by a center insulator. Wire dimen-

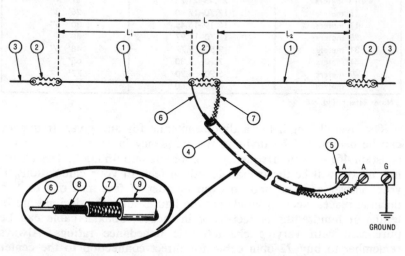

**Fig. 4-6. Dipole antenna construction details.**

## Table 4-1. Dimensions Required for a
## Single Dipole Antenna Shown in Fig. 4-6

| Frequency Band | Antenna Dimensions | | |
|---|---|---|---|
| | L1 | L2 | Length |
| 80 meters | 58' 10" | 58' 10" | 117' 8" |
| 40 meters | 33' 0" | 33' 0" | 66' 0" |
| 31 meters | 23' 5" | 23' 5" | 46' 10" |
| 25 meters | 20' 0" | 20' 0" | 40' 0" |
| 19 meters | 16' 0" | 16' 0" | 32' 0" |
| 15 meters | 11' 0" | 11' 0" | 22' 0" |
| 10 meters | 8' 4" | 8' 4" | 16' 8" |

sions required are listed in Table 4-1. The length of the antenna between the terminal insulators is about 5 percent less than half the wavelength to which it is tuned. Or, in other words, a half-wave antenna for the band in use is a resonant wire cut 5 percent shorter than one half the wavelength of the center frequency of the band. The international shortwave broadcast bands are not very wide and a resonant wire cut for the center of a given band will be down very little in amplitude on the ends of the frequency dial. Antenna lengths for the international shortwave bands are listed in Table 4-2.

## Table 4-2. Antenna Lengths for International
## Shortwave Bands

| Band | Wavelength (MHz) | Antenna Length |
|---|---|---|
| 13 meters | 21.45–21.75 | 21' 9" |
| 16 meters | 17.70–17.90 | 26' 3" |
| 17 meters | 15.10–15.45 | 30' 5" |
| 25 meters | 11.70–11.97 | 40' |
| 31 meters | 9.50– 9.77 | 48' 6" |
| 41 meters | 7.10– 7.30 | 65' |
| 49 meters | 5.90– 6.20 | 77' |
| 60 meters | 4.75– 5.60 | 95' |

Note: Refer to Fig. 4-6.

The overall length for a dipole antenna for any given frequency can be determined by dividing the frequency in megahertz into the constant 468—or the frequency in kilohertz into 468,000. The resulting answer will be the overall length in feet. In your calculations, if your answer goes beyond an even number of feet into one or two decimal places, keep in mind that the figures after the decimal are in tenths or hundredths of feet—not in inches. Coaxial cable can be purchased with varying characteristic impedance ratings. Always remember to buy 72-ohm cable for direct connection to the center of a single dipole.

The numbers in the following construction details correspond to those numbers called out in Fig. 4-6. You may want to refer to them as you read. Your antenna can be cut or made to a length for optimum operation on an often-used frequency or specific shortwave band. The dipole antenna is a wire of specific length (1) with an insulator connected at each end and also in the center (2). The insulators at the extremes are attached to ropes that can be tied to a tree or post. A two-wire feeder line (4) is required for connection to your receiver (5). The center conductor (6) and the wire braid (7) should be connected as shown. Recall that 468 divided by the frequency equals the antenna length in feet. The two-wire feeder line (4) is a 72-ohm coaxial cable. No. 8 is insulation and No. 9 is the outside protective insulation covering.

A dipole antenna will also work well at three times the frequency for which it was made. An antenna cut for the 6-MHz shortwave band will also work well at 18 MHz, the upper end of the 16-meter band.

Dipoles can be made that will operate on several bands by adding "wave traps" like those shown in Fig. 4-7 at various points along the

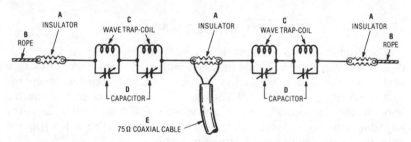

Fig. 4-7. Dipole antenna with wave traps allows multiband reception.

length of the antenna. The purpose of these parallel resonant circuits is to act as insulators for their designed resonant frequencies and as conductors for other frequencies. You can achieve a wave trap by using home-made coils of about 1½- to 2-inches diameter and spacing the turns according to the size of the wire. The diameter has to be substantial—about 2 mm (AWG12). A trap can be constructed as shown in Fig. 4-8. It must be protected from the weather by a plastic cover. The trimmer-capacitor, connected parallel to the coil, can be housed inside it along with the strain insulator that separates the antenna portions, The most difficult thing for you to do will be to properly tune the circuit to the desired frequency, which is the middle of the shortwave broadcast bands. To accomplish this, use a frequency meter in combination with a vacuum-tube voltmeter. Construction of a trap dipole antenna might best be illustrated with the following

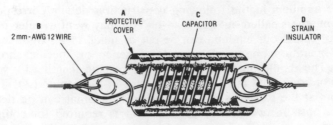

Fig. 4-8. Wave trap construction details.

example. Suppose that you want to make the antenna resonant at 15.3, 17.8 and 21.6 MHz for optimum performance in the 13-, 16-, and 19-meter bands. The total length of the antenna—between the terminal insulators—then becomes 30 feet, the length of the 16-meter band at 26 feet, and the 13-meter band at 21 feet. The traps are inserted at the 26- and 21-foot points to act as insulators—without interfering with the signals of resonance—you will have a single dipole antenna that is resonant at three different frequencies. The impedance of this antenna is 73 ohms.

## FOLDED DIPOLE ANTENNA

The folded dipole antenna is an improved version of the single dipole described above. This antenna will work over a greater range of frequencies and it will effect a more efficient transfer of energy to the receiver. As Fig. 4-9 shows, it employs two end-connected wires in the flat top. The wires are spaced about an inch apart—the impedance has been increased to nearly 300 ohms. Table 4-3 lists the antenna length for various frequency bands.

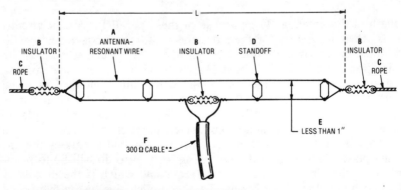

*300-OHM RIBBON LINE (TV RIBBON, AMPHENOL 214-559 OR EQUIVALENT)
**-TV RIBBON, AMPHENOL 214-056 OR EQUIVALENT.

Fig. 4-9. Folded dipole antenna offers wide frequency range.

**Table 4-3. Antenna Dimensions for a Folded Dipole**

| Frequency Band | Antenna Length |
|---|---|
| 15 meters | 22' 0" |
| 19 meters | 32' 0" |
| 25 meters | 40' 0" |
| 31 meters | 46' 6" |
| 40 meters | 65' 6" |
| 49 meters | 78' 0" |
| 60 meters | 97' 0" |
| 90 meters | 142' 0" |

## TRIPLE DIPOLE ANTENNA

Combining three dipoles into one antenna creates an all-wave antenna system that can cover the frequency range from 5 MHz to 30 MHz. It is composed of three separate dipoles each connected to the other and to the transmission wire at the center of the antenna (see Fig. 4-10). The dipole wires are fanned away from each

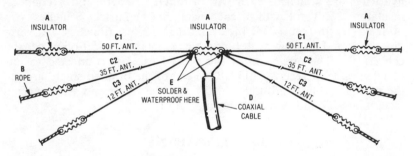

**Fig. 4-10. Triple dipole antenna construction details.**

other. To achieve optimum results over the entire 5-MHz to 30-MHz frequency range, a 50- or 75-ohm coaxial cable should be used for the lead-in wire. Use hard-drawn copper wire or steel-core wire for the top sections of the antenna—this will add strength to the overall construction. Solder and waterproof the connections where the three antennas join on each side of the center insulator.

## FAN VERTICAL ANTENNA

The fan vertical antenna shown in Fig. 4-11 is a multiple-band receiving antenna which has the same basic construction as the long-wire antenna in Fig. 4-1. The numbers in the following construction details correspond to the numbers called out in Fig. 4-11. You may want to refer to them as you read. A rope (1) is attached to a 34-

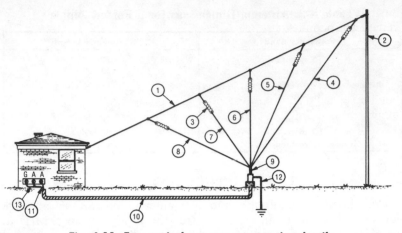

Fig. 4-11. Fan vertical antenna construction details.

foot mast (2)—the height need only be approximate. To the rope, five wires are attached, each with a different length and a connect-insulator (3). The length of each wire is: (4) 25 feet 6 inches, (5) 20 feet, (6) 15 feet, (7) 10 feet, and (8) 12 feet 6 inches. All wires are connected to the center conductor (9) of a 50-ohm coaxial cable (10). The other end of the coaxial cable is connected to terminal A on your receiver (11). One end of the braided shield around the coaxial cable is grounded (12) and the other end is connected to terminals G and A of the receiver (13).

## BEAM ANTENNA

This antenna can be used with a rotating motor since the directional sensitivity is not the same around a 360-degree circle. You can turn the rotor so that the antenna can be pointed toward an incoming signal for best reception (see Fig. 4-12). The beam

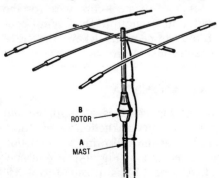

Fig. 4-12. Commercial beam antenna has directional rotor.

antenna also has more gain than a dipole antenna and signals received will be much stronger.

## INDOOR ANTENNA

When you live in an apartment or school dormitory where the landlord takes a dim view of antennas coming out from windows or mounted on top of the roof, you might consider an indoor antenna. This kind of antenna can be constructed in different ways.

One method is to use a large expanse of aluminum foil along one wall or more, as shown in Fig. 4-13, with the lead-in wire connected

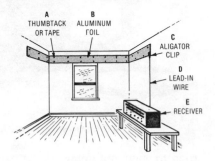

Fig. 4-13. Aluminum foil attached to the walls can provide a workable indoor receiving antenna.

to it by tape or an alligator clip. Another method is to use a metal bed frame as an antenna. These may not be ideal solutions, but they have proved effective.

An antenna system that has been recently introduced may solve the problems of apartment dwellers and those persons who are not able to put up a dipole or long-wire antenna. (See Fig. 4-14.) This is an "all-wave receiving antenna" that consists of an interior module —styled in textured black with genuine teakwood sides—and an exterior module that includes an antenna rod and 50 feet of RG58, 50 ohm coaxial cable. For best performance the exterior module should be mounted as high as possible. The author mounted the module against the top of the garage at about 10 feet in the air. When connected to a receiver, stations from all over the world were "pulled in." No high masts or other complicated exterior antenna systems are needed when using this indoor all-wave antenna.

A final note. The subject of DX antennas cannot be dealt with completely within this limited number of pages. The variety of antenna forms and properties is so diverse that many books have been written on the subject. It is hoped, however, that what you have been exposed to thus far will stimulate you to experiment with various antennas. It is recommended that you read more about them, too, in

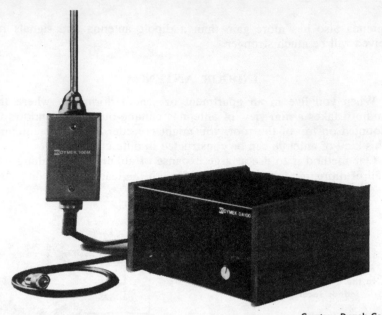

Courtesy Dymek Corp.

**Fig. 4-14. All-wave receiving antenna for indoor use.**

order to gain optimum results from your equipment. When compared to the cost of a receiver, those for the antenna are relatively small and it will be worth your while to spend a little more, perhaps, in order to achieve the best results.

# CHAPTER 5

# QSL Reports and
# Where to Send for Them

Now that you are set up to listen to all those exotic sounds, words, and music from far-away places, what is it that you will hear? Often you will hear different codes being used; and in different languages.

And what do you do after you have heard an exciting broadcast from a radio station on the other side of the world?

In this chapter we will look at those codes and will also discuss QSLs and how they can be acquired.

## GETTING THOSE QSLs

Whatever you call them—QSLs, confirmations, verifications, or veries—the collecting of these letters or cards from the stations you hear, stating that you actually did hear them, is an added part of the fun of DXing. It is one thing to know that you heard a station. It is even more fun to have a postcard or letter from the station specifying that your reception report was correct. Often QSLs are colorful, attractive cards like the samples shown in Fig. 5-1 which you can mount on the wall of your DX station or place in an album. They are tangible evidence of your hobby that you can show to friends and relatives. You can tell others that you have been able to hear 25 or 50 or 100 different countries. But being able to show them responses from the stations you have heard is even more impressive.

Getting a QSL is something of an art in itself, sometimes it can be as hard as actually hearing the station. While many of the large, international shortwave broadcasters, such as Radio Sweden shown

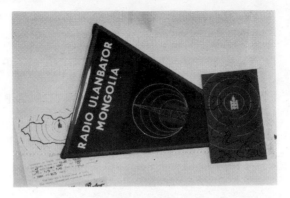

*(A) Mongolia.*

*(B) Sweden.*

*(C) Ghana.*

**Fig. 5-1. Sample QSL reports.**

in Figs. 5-2 through 5-5, are happy to verify listeners reports, the smaller local and regional sw stations sometimes are less apt to reply with QSLs. In some cases it seems to be a lack of interest. After all, they are most interested in reaching home audiences and if they are heard by overseas listeners, it is, at best, a novel event. More often, therefore, smaller stations are slower to send a QSL because they do not have the staff to answer the mail from SWLs, or they really do not know much about the hobby and what a QSL is, much less why you want it.

Fig. 5-2. Radio Sweden—Hörby transmitting station.

No broadcast station is under obligation to send you a QSL. The staff does so out of courtesy, or because your report has helped the station engineers determine transmission patterns and actual interference from other transmitters. Therefore, *politeness* in your reception report is essential. Report your reception and at the end of your letter ask—DO NOT DEMAND—that your report be verified as correct. The right attitude and approach often makes a great deal of difference.

## RECEPTION REPORTS

A reception report is usually classified as being in one of four categories: (1) from those who listen for pleasure and entertainment only; (2) from those who listen to gain a better understanding of other countries or to learn a foreign language; (3) from those who listen for the purpose of reporting and collecting verifications and QSL cards; and (4) from those who listen and monitor for the pur-

**Fig. 5-3. One of three 500-kW shortwave transmitters used by Radio Sweden.**

pose of reporting technical data of value to the engineering department of the station heard. A good listener can qualify in each category, if he so desires.

When reporting for verifications, it is important for you to remember that it is necessary to report reception over a period of time—at least one-half hour, when possible. The listening time period can, of course, be adjusted as conditions and transmission time of the station dictate. Longer reporting periods are preferred, and, accordingly, are much more useful to the station, particularly if the report contains pertinent and technical information. It is not considered unusual to monitor and report on a reception over a period of one, or even two, hours during the transmission time of the station. However, many station schedules may run for a shorter period of time—therefore, you must judge for yourself how long you can monitor the transmission of any particular station.

### Writing the Report

There are four basic parts to a reception report that you write to a particular station you hear. First, to establish that you actually did hear that station, you must tell *what* you heard. Second, for your report to be of any use, tell the station *how well* you heard its broadcast. Next, the station normally is interested in you, the equipment you used to hear the station and what you thought of its programs,

**Fig. 5-4. One of three shortwave antennas used by Radio Sweden.**

your likes and dislikes. Finally, you get to request for a QSL letter or card.

The last two parts of your report are self-explanatory. However, a few words about *what* you heard and *how well* are in order. To allow the station to check its program log and determine that you did, in fact, hear its broadcasts, you have to include information on date, time, and frequency of reception. It is important to list the frequency on which you heard the station. Some stations use a number of different frequencies in the same meter band. If your receiver has accurate frequency readout, knowing the correct frequency in kilohertz is no problem. If it does not, you will have to listen carefully for the announced frequency or, as a last resort, make the best "guesstimate" you can from your dial. And, if you are not sure of the frequency, be sure to note that fact in your reception report.

Fig. 5-5. Shortwave broadcast console in the studio at Radio Sweden.

The time used in your report should be Greenwich Mean Time (GMT) or the local time at the station site. Generally, GMT is easier to calculate and most of the major stations understand it. GMT is the time at the zero meridian of longitude, which runs through Greenwich, England. GMT is five hours ahead of the U.S. Eastern Standard Time, six hours ahead of Central Standard Time, seven hours ahead of Mountain Standard Time and eight hours ahead of Pacific Standard Time.

To prevent confusion with A.M. and P.M. references, generally the 24-hour clock is used. Under this system of reference, midnight is expressed as 0000, 1 A.M. as 0100, and so forth. Noon is 1200, 1 P.M. becomes 1300, 2 P.M. is 1400. Thus, 4:45 A.M. is expressed as 0445; 15 minutes past noon is 1215 and ten minutes before midnight becomes 2350.

It is necessary, too, to tell the station in your reception report what you heard in the way of programming. Be as detailed as you can.

Note specific times when you heard certain programs or announcements. Mention program names and any other information that will help the station determine that you were really listening to it and not some other broadcaster. That means that vague program details, such as "0230 GMT, man talking; 0233 GMT, music;" just are not good enough.

## How You Heard It

Along with the program details—time, date, and frequency—you should report to the station just how well it was received. This will assist the station in future program planning to adjust schedules and frequencies if reception is unsatisfactory over a period of time in the listener's area. It also allows the station engineering staff to have more knowledge of what the transmitter signal is doing and where improvements can be made. You, the DXer, will benefit by getting better programming and better overall reception.

You may include comments on the programming heard, what you liked and what you did not like. But remember to be constructive in your criticism. Do not unjustly criticize the program without some thought of offering suggestions of preference, rather than a particular program transmitted. Each shortwave broadcasting station serves thousands, if not millions, of listeners and it is quite impossible to suit the ears of every person who tunes in its programs. Constructive criticism and suggestions as to the preference of the listener are welcomed by the station and enable the program planners and staff personnel to plan for future programs that will be most enjoyed by the majority of their listening audience.

When preparing the actual reports, list all information in letter form. Do not send reports on postcards. They simply cannot accommodate enough of the necessary information to be of value to the station. If you have your own SWL card, you could send it along with your report. But do not use your card strictly for reception reporting purposes.

# REGIONAL BROADCASTERS

The prize of every QSL collector: the low-powered regional broadcaster. Those are the stations that direct their broadcasting to listeners inside their country or close by. Thus, the engineer of an African, Latin American, or Asian station probably could care very little how his station is being received in the U.S. or Canada. Since a report from you as a North American DXer is of no real use to him, he would feel no particular obligation to answer you with a verification. The first thing to remember, therefore, is that any reply you receive from a regional station is a personal favor from them to you.

The following are a few general ideas on how to report to regional broadcasters:

(1) Unless the station is known to answer reports submitted in English, send your report in the language of the country in which the station is located. Since these stations have no foreign service, they probably would not have a staff capable of handling foreign-language letters. If you do not happen to be a linguist, form letters in French, Spanish, Portuguese, and other languages can be obtained from several radio clubs.

(2) If it is at all possible, send mint stamps of the country in which the station is located. These are much more convenient than IRCs which have to be taken to a post office to be exchanged—not always "just around the corner." The major North American supplier of mint stamps in the correct units and at quite reasonable prices is: DX Stamp Service (Mr. G. N. Robertson), 83 Reder Parkway, Ontario, NY 14519, U.S.A.

(3) Many regional stations are commercially operated. In your program details, try to include the brand names of the sponsor's products that you hear advertised. This will help convince them that your report is valid.

(4) Include with your report postcards of your home town, the U.S., cancelled stamps, newspaper pictures, etc. An envelope filled with postcards and other items is likely to receive more attention than one that contains only a single letter; you are seeking to have your envelope opened and your letter read.

(5) Write all your reports promptly after reception and send them by air mail.

Eventually, you will run across a station that does not answer your report within a reasonable length of time. In that event, you can do either of two things: If you receive the station from time to time, send them a completely new report; or if the station is hard to hear, send them a follow-up report with a letter explaining that you have not received a verification on your first report.

There are some stations that answer few, if any, reception reports. The only thing to do is to keep writing and hope that yours will be the one to cause them to change their mind.

## REPORTING CODES

Reporting codes are very useful to you as well as to the station receiving the report. You will find this method of reporting to be simple, yet comprehensive. Through the years there have been a number of reporting codes in use, but of all of them perhaps the best known and most widely used are the "RST code," the "555 code," and the "SINPO code."

## RST Code

The RST code reads as follows:

R (*Readability*)
1—unreadable
2—barely readable, occasional words distinguishable
3—readable with considerable difficulty
4—readable with practically no difficulty
5—perfectly readable

S (*Signal Strength*)
1—faint, signals barely perceptible
2—very weak signals
3—weak signals
4—fair signals
5—fairly good signals
6—good signals
7—moderately strong signals
8—strong signals
9—extremely strong signals

T (*Tone*)
1—sixty-cycle ac or less, very rough and broad
2—very rough ac, very harsh and broad
3—rough ac tone, rectified but not filtered
4—rough note, some trace of filtering
5—filtered rectified ac, but strongly ripple-modulated
6—filtered tone, definite trace of ripple modulation
7—near pure tone, trace of ripple modulation
8—near perfect tone, slight trace of modulation
9—perfect tone, no trace of ripple modulation of any kind

## 555 Code

The 555 and SINPO codes have proven to be the most widely used as well as the most popular with SWL DXers. The 555 code is as follows:

*Signal Strength* (*QSA*)
5—excellent
4—very good
3—good
2—fair
1—poor
0—inaudible

*Interference* (*QRM*)
5—none
4—slight

3—moderate
2—severe
1—extreme
0—total

*Overall Merit* (*QRK*)
5—excellent
4—very good
3—good
2—fair
1—poor
0—unusable

## SINPO Code

The SINPO code is intended as a means of representing a fairly accurate and nontechnical evaluation of received radio signal quality. The acronym SINPO is **S**ignal Strength, **I**nterference, **N**oise, **P**ropagation, and **O**verall merit. For you as a DXer submitting an unsolicited reception report, the responsibility for sending useful information lies in fairly and accurately relating conditions of a particular broadcast.

*S* (*Signal Strength*)—This pertains to the amount of radio signal energy actually captured by the receiving station. The one to five scale offers the possibility of rating the signal strength from almost none to very strong.

5—very strong
4—good (strong)
3—fair
2—weak (poor)
1—very weak (barely audible)

*I* (*Interference*)—This refers to man-made interference from other radio transmissions. The increasing problem of more and more stations operating within allocated limits makes this particular observation quite important, especially when more than one geographic area is the intended target of a certain broadcast. When interference is observed, information regarding the identity of the interfering station and its operating frequency is of considerable value and should be noted as a supplement to your SINPO evaluation.

5—none
4—slight
3—moderate
2—severe
1—extreme

*N* (*Noise*)—This refers to the natural atmospheric, electrostatic noise. If you wish to contribute supplemental information of some value, a reference to the normal conditions for the particular frequency band, season, and time of day can be included, provided that you are familiar with current reception conditions for the particular band.

5—none
4—slight
3—moderate
2—severe
1—extreme

*P* (*Propagation*)—This is perhaps the most confusing and misinterpreted element to the SINPO code. Over long distances, the fact that the program is heard at all certainly suggests that at least moderately favorable propagation conditions exist and confirms the technician's calculation. Therefore, you should consider only the degrading characteristics of the signal itself—is it fading.

5—none
4—slight
3—moderate
2—severe
1—extreme

NOTE: Fading is much misunderstood; it refers to variations in signal strength which are almost always present during shortwave reception. It can be any of the following:

Rhythmic: regular recurring variations in signal strength—starts out slow and gradually increases in rapidity.

Random: irregular changes—most often experienced by DXers.

Selective: distortion of the modulation by different fading effects on the sidebands of a signal. Voices sound as if they are being varied in pitch.

Auroral: fluttering sound as signals pass through or near the polar auroral zones. More noticeable during the winter in the northern hemisphere.

*O* (*Overall Merit*)—This is a summation of the general readability of the transmission. It is the combined net result of the other four parts of the SINPO code. When evaluating overall merit, you should consider that you have only a single choice from the five options with which to describe the value of the transmission to the broadcaster. After all, the broadcaster is interested in just how well the program is being heard by his regular audience.

5—excellent
4—good

3—fair
2—poor
1—unusable

To a large, international broadcaster, the notation "SINPO 34323" is usually sufficient. To smaller, regional broadcasters you should expand the above notation to:

Signal strength . . . . . . fair
Interference . . . . . . . . slight
Noise . . . . . . . . . . . . . moderate
Fading . . . . . . . . . . . . severe
Overall Merit . . . . . . . understandable

## FOREIGN-LANGUAGE REPORTS

Besides using the English language in your reports, think of using the language of the country you will send your report to. Sure, around the world, we recognize English as the universal language used or, at least, spoken by most educated persons. English is well received by a majority of stations and is readily understood by station personnel for purposes of verification of reception reports. However, there will come a point in your hobby when the QSLs are no longer so easy to come by. The reason is usually because of the lack of communication. One of the DX clubs in this country (listed later in this chapter), the NASWA, employed the services of several knowledgeable linguists in its membership to develop several foreign-language report guide forms. Initially, the French, Spanish, and Portuguese languages were developed for use because many of the areas of the world not responding for QSL purposes used those languages. Most of western Africa uses French. Some of the West Indies and all of Latin America (excluding Brazil) use Spanish. Portuguese can be found as a native tongue in many areas of Africa and Asia. These language report forms were a boon and many QSL totals began to climb.

Later on, forms in the Indonesian language were devised. From other clubs and from other NASWA associates in Europe came forms developed in the Russian and Danish languages. With the help of NASWA members, more forms were developed for the exotic languages of Farsi, Urdu, Persian, Vietnamese, and Chinese. Another language—Arabic—was added through the courtesy of a German DX club. NASWA members now have the ability at least to communicate their basic intent and request to all the users of these various languages. This club continually strives to improve the development of new and more efficient guides.

# IDENTIFYING A FOREIGN LANGUAGE

Even if the broadcast is not in English, you can often very easily understand what is being said. Through a process of elimination you can check the most recent bulletin loggings and the station guide to ascertain who may reasonably be expected to occupy the frequency you are tuned to. Consider each one carefully. What is the time of day? What is the frequency used?

Next, you should consider whether the station is supposed to be on the air at the time period it is heard. Also, try to determine whether the broadcast is regional or international in scope.

Consider the language. Can you identify it? A low-frequency station below 49 meters is likely to use the languages indigenous within its own borders. How do you identify, for example, a very weak station broadcasting in Spanish when the station guide has at least four Latin countries listed in that band. It is not easy, but it can be done. One method you can use is to tune to a few other known stations near the same frequency and note which regions you are hearing best at that moment. One broad generality, for example, is that most Venezuelans use frequencies ending in zero. Columbians mainly use frequencies ending in five. One interesting fact is that, by international treaty, no commercial broadcasting can be done on the 41-meter band in the western hemisphere—the one exception to the rule is Canada.

There is another logic: very powerful broadcasts come, most likely, from the very large stations—and they may use any language. On frequencies of 31 meters and above, the most powerful transmissions heard are likely to be from one of the following: Voice of America, Moscow, Tirana, BBC, Deutsche Welle (West Germany), Netherlands, Cairo, or Saudi Arabia. As a beginning DXer you will be well ahead if you devise several surveys yourself to aid in identifying the different stations. There are many ways to acquire the ability to recognize the different languages—the most important one is *experience*.

One way is to listen to the Voice of America or BBC. They announce in English which language is going to be used in the next broadcast. Also, different stations use particular identifications, such as "Here is . . ." or "Speaking . . ." or "This is . . . ."

## Spanish and Portuguese

Spanish has five cardinal vowels, a, e, i, o, u—and all spoken clearly, sharply and without nazalization. Castillian Spanish has the sound of English "th" in cinco (thinko), cenar (thenar)—and "ly" in llamar (lyamar). In Latin America they pronounce these words 'sinko,' 'senar,' and 'yamar,'

Portuguese has a total different vowel system with many nazalized sounds. Final 'a' is pronounced 'e' in Portugal, final 'o' as 'u.' Typical endings are nazalized cao, im and em. "Here is" . . . is "aqui" in Spanish—pronounced Aki and Eki in Portuguese. "This is" . . . is "esta es" in Spanish and "fala" in Portuguese.

### Arabic

There are three types of Arabic. Egyptian and Sudanese Arabic are totally unlike the others in one important aspect—the sound of "g" as in "game" is heard in "gumhuryah" . . . whereas in Iraq, Saudi Arabia, and others, the "g" is pronounced as in "gem."

### Pacific Islands

Polynesian languages have fewer different sounds than any other language. They have the five cardinal vowels: a, e, i, o, and u; and only about seven or eight consonants, mainly h, k, l, r, m, and n.

### Far Eastern Languages

These consist of tonal and nontonal languages. If it is tonal, it must be Chinese, Vietnamese, Thai, Laotian, Cambodian, or Burmese. These are all languages of mainland Asia. If the language is not so much tonal, then it is probably Japanese, Korean, Philippine, or Malaysian.

All Japanese words end in a vowel or n. Korean may resemble Japanese but it is more breathy, especially the consonants. Chinese is a very tonal language, changing word meanings as much as four or five times depending on the pitch in which it is spoken in relation to other words.

### Major Languages of the World

| *Arabic* | *English* | *French* |
|---|---|---|
| Algeria | United Arab Emirates | New Zealand |
| Bahrain | Zanzibar | Nigeria |
| Egypt | | Philippines |
| Iraq | *English* | Seychelles |
| Jordan | Australia | Singapore |
| Kuwait | Barbados | United States |
| Lebanon | Bermuda | Windward Islands |
| Libya | Canada | |
| Morocco | Fiji | *French* |
| Quatar | Ghana | Afars & Isas |
| Saudi Arabia | Gt. Britain | Algeria |
| Southern Yemen | Guyana | Belgium |
| Sudan | Hong Kong | Burundi |
| Syria | India | Cambodia |
| | Jamaica | Cameroun |

| | | |
|---|---|---|
| Cent. African Rep. | Azores | Philippines |
| Chad | Cape Verde Is. | Spain |
| Comoro Is. | Brazil | Venezuela |
| Congo | Guinea | Uruguay |
| Dahomey | Portugal | |
| France | Sao Tome | *German* |
| Gabon | | |
| Guadeloupe | *Spanish* | Germany, East |
| Haiti | Argentina | Germany, West |
| Ivory Coast | Bolivia | Austria |
| Laos | Canary Is. | Switzerland |
| Malagasy | Chile | |
| Mali | Colombia | *Chinese* |
| Martinique | Costa Rica | China |
| Mauretania | Cuba | Hong Kong |
| New Caledonia | Dominican Rep. | Singapore |
| Reunion Is. | Equador | Taiwan |
| Rwanda | El Salvador | |
| Senegal | Guatemala | |
| Tahiti | Honduras | *Swahili* |
| Upper Volta | Mexico | Burundi |
| Zaire | Nicaragua | Kenya |
| | Panama | Rwanda |
| *Portuguese* | Paraguay | Somali Rep. |
| Angola | Peru | Tanzania |

Who are these broadcasters talking in these different languages and where are they located? The rest of this chapter should help you find the answers to these interesting questions, as well as many others in your DX adventures.

## INTERNATIONAL BROADCASTERS—
## WHERE TO SEND RECEPTION REPORTS

When you are tuned in to a shortwave station you will probably want to send a reception report in the hope of receiving a QSL card. The following are mailing addresses for some of the interna- tional stations. *Shortwave Listener's Guide* lists stations broadcasting from more than 100 countries all over the world. Stations are listed alphabetically by country and by frequency. This book is published by Howard W. Sams & Co., Inc., 4300 West 62nd Street, Indianap- olis, IN 46268 ($4.95). Additional addresses can be found in *The World TV Radio Handbook* and *SWL Address Book*—both available from Gilfer Assoc., P.O. Box 239, Park Ridge, NJ 07656.

| | |
|---|---|
| Afghanistan | Radio Afghanistan, PO Box 554, Kabul |
| Albania | Radio Tirana, Rue Ismail Quemal, Tirana |

| | |
|---|---|
| Algeria | Radio Diffusion Television Algerienne, 21 Blvd. des Martyrs, Algiers |
| Angola | Emissora Oficial, C.P. 6455, Luanda |
| Argentina | Radiodifusora Argentina al Exterior (RAE), Sarmineto 151, Buenos Aires |
| Ascension Is. | BBC Atlantic Relay Station, c/o BBC, P.O. Box 76, London WC 2B 4PH, England |
| Australia | Radio Australia, Box 4826, G.P.O., Melbourne |
| Austria | Austrian Radio, Techn. Dept., P.O. Box 200, A-1043 Vienna |
| Bangladesh | Radii Bangladesh, 20 Green Rd., Dacca 5 |
| Belgium | Belgian Radio & TV, Intern. Service QSL Bureau, P.O. Box 26, B-1000 Brussels |
| Bolivia | La Cruz del Sur, C.P. 75, Cajon 1408, La Paz |
| Botswana | Radio Botswana, P.O. Box 52, Gaberones |
| Brazil | Radio Nacional de Brasilia, P.O. Box 1620, Brasilia |
| Bulgaria | Radio Sofia, 4 Bd. Dragan Tsankov, Sofia |
| Burma | Burma Broadcasting Service, Prome Rd., Kamayut P.O., Rangoon |
| Cameroun | Radio Yaounde, Boite Postale 281, Yaounde |
| Canada | Radio Canada Intern., P.O. Box 6000, Montreal H3C 3A8 |
| Canary Is. | Radio Nacional de Espana, Centro Emisor de Atlantico, Valentin Sanz 4, Pisos 5 y 6, Santa Cruz de Tenerife |
| Cent. Afr. Rep. | Radio Bangui, Techn. Services Dept., Radiodiffusion Nationale Centrafricaine, B.P. 950, Bangui |
| Chile | La Voz de Chile, Casilia 13130, Santiago |
| China | Radio Peking, Peking |
| China (Taiwan) | Voice of Free China, Broadcasting Corp. of China, Overseas Dept., 53 Sec. 111, Jen Ai Rd., Taipei |
| Colombia | Radio Nacional de Colombia, Apt. Aereo 51-72, Bogota |
| Costa Rica | TIFC, Faro Del Caribe, Apartado 287, San Jose |
| Cuba | Radio Havana, P.O. Box 7026, Havana |
| Czechoslovakia | Radio Prague, Prague 2 |
| Denmark | Radio Denmark, TV-Byen, DK-2860 Soborg, Copenhagen |
| Ecuador | Voice of the Andes, HCJB, Casilla 691, Quito |
| Egypt | Radio Cairo, P.O. Box 566, Cairo |
| Ethiopia | Radio Voice of the Gospel, ETLF, P.O. Box 654, Addis Ababa |
| Finland | Radio Finland, Foreign Relations Dept., Finnish Broadcasting Co., Helsinki 26 |
| Germany, East | Radio Berlin Intern., Nalepastrasze 18-50, Berlin 116 |
| Germany, West | Deutsche Welle, Voice of Germany, P.O. Box 344, Cologne 5 |
| Ghana | Radio Ghana, P.O. Box 1633, Accra |
| Great Britain | BBC, Box 76, Bush House, Strand, London |
| Guatemala | Radio Cultura, Apartado 601, Guatemala City |
| Haiti | Radio 4VEH, La Voix Evangelique, B.P. 1, Cap Haitien |
| Honduras | HRVC, La Voz Evangelica, Apartado 270, Tegucigalpa |

| | |
|---|---|
| Hungary | Radio Budapest, Brody Sandor 5-7, Budapest VIII |
| India | All India Radio, External Services, Box 500, New Delhi 1 |
| Indonesia | Voice of Indonesia, Box 157, Jakarta |
| Israel | Israel Broadcasting Authority, Box 1082, Jerusalem |
| Italy | Italian Radio and TV (RAI), Caselle Postale 320, Rome |
| Ivory Coast | Radio Abidjan, B.P. 2261, Abidjan |
| Japan | Radio Japan, Jinnan 2-2-1, Shibuya-ku, Tokyo 100 |
| Korea, North | Radio Pyongyang, Pyongyang, Dem. People's Rep. |
| Korea, South | Radio Korea, 8 Yejangdong Joong-yu, Seoul, Rep. Korea |
| Kuwait | Kuwait Broadcasting & TV Service, P.O. Box 397, Kuwait |
| Lebanon | Radio Lebanon, Ministery of Information, Beirut |
| Liberia | Radio Station ELWA, Box 192, Monrovia |
| Malaysia | Radio Malaysia, Federal House, P.O. Box 1074, Kuala Lumpur |
| Mali | Radiodiffusion du Mali, B.P. 171, Bamako |
| Mexico | Radio Mexico, XERMX, Apartado 20100, Mexico 20 DF |
| Mongolia | Radio Ulan Bator, CPO Box 365, Ulan Bator, Mongolian People Republic |
| Nepal | Radio Nepal, Dept. of Broadcasting, Katmandu |
| Netherlands | Radio Nederland, P.O. Box 222, Hilversum |
| New Caledonia | Radio Noumea, B.P. 327, Noumea |
| New Zealand | Radio New Zealand, P.O. Box 2396, Wellington |
| Nigeria | Voice of Nigeria, Broadcasting House, Lagos |
| Norway | Radio Norway, Bj. Bjornsons Plass 1, Oslo |
| Pakistan | Radio Pakistan, 71 Garden Rd., Karachi |
| Papua New Guinea | National Broadcasting Commission, Box 1359, Boroko |
| Peru | Radio Nacional, Av. Peiti Thouars 441, Lima |
| Philippines | Far East Broadcasting Co., Box 2041, Manila |
| Poland | Polish Radio, P.O. Box 46, Warsaw |
| Portugal | Radio Portugal, Rue de Quelhas 21, Lisbon |
| Rhodesia | Rhodesian Broadcasting Corp., P.O. Box 2696, Salisbury |
| Rumania | Radio Bucharest, P.O. Box 111, Bucharest |
| Russia | Radio Moscow, U.S.S.R. |
| Saudi Arabia | Broadcasting Service of the Kingdom of Saudi Arabia, Box 2476, Riyadh |
| Sierra Leone | Sierra Leone Broadcasting Service, New England, Freetown |
| Singapore | Radio Singapore, Dept. of Broadcasting, Ministry of Culture, P.O. Box 1902, Singapore |
| South Africa | Radio RSA, Box 4559, Johannesburg |
| Spain | Radio Nacional de Espana, General Yague 1, Madrid 20 |
| Sri Lanka | Sri Lanka Broadcasting Corp., P.O. Box 574, Colombo 7 |
| Sweden | Radio Sweden, S-10510, Stockholm 1 |
| Switzerland | Swiss Broadcasting Corp., Overseas Service, CH3000, Berne 16 |
| Syria | Radio Damascus, Omayad Square, Damascus |

| | |
|---|---|
| Tahiti | Radio Tahiti, B.P. 125, Papeete |
| Tanzania | Radio Tanzania, P.O. Box 1178, Dar es Salaam |
| Togo | Radiodiffusion de Togo, B.P. 434, Lome |
| Turkey | Radio Ankara, General Mudurlugo, Mithat Pasa Caddesi 37, Ankara |
| Uganda | Radio Uganda, P.O. Box 2038, Kampala |
| Ukranian U.S.S.R. | Radio Kiev, Radio Centre, 26 Kreschatik, Kiev, U.S.S.R. |
| U.S.A. | Voice of America, c/o Frequency Div., Washington, D.C. 20547. |
| U.S. Armed Forces | Armed Forces Radio & TV Service, AFRTS, Washington, D.C. 20305 |
| Vatican | Vatican Radio, Vatican State |
| Venezuela | Radio Rumbos, Apartado 2618, Caracas |
| Vietnam | Voice of Vietnam, 58 Quan-su Str., Hanoi |
| Windward Is. | Radio Grenada, P.O. Box 34, St. George's, Grenada |
| Yugoslavia | Radio Belgrade, External Broadcasting, 2 Hilendarska, Belgrade |
| Zaire | La Voix du Zaire, B.P. 3171, Kinshasa |
| Zambia | Zambia Broadcasting Service, P.O. Box RW 15, Lusaka |

## DX CLUBS—DOMESTIC AND INTERNATIONAL

There are quite a number of domestic and international clubs, providing the DXer with miscellaneous services, such as pamphlets on different subjects; listings of stations; club activities; etc. At a nominal cost, most of the clubs will send the new DXer a sample of their bulletin. You, then, have to determine for yourself which club comes closest to your own area of DXing, i.e., Broadcast Band, Shortwave, etc.

### Domestic Clubs

*ANARC*—Association of North American Radio Clubs; 557 N. Madison Ave., Pasadena, CA 91101 . . . This is a confederation of DX radio clubs—it is not a club in itself. Send for information. Members are the following.

*ASWLC*—American Shortwave Listeners Club; 16182 Ballad Lane, Huntington Beach, CA 92649 . . . Shortwave, broadcast band, utilities, QSL, SWL. Monthly bulletin. Sample bulletin 75 cents. Dues $12.00/year. Founded 1959.

*CIDX*—Canadian International DX Radio Club; 169 Grandview Ave., Winnipeg, Manitoba, Canada R26 OL4 . . . Shortwave and Broadcast bands, Utility and Ham bands. Monthly bulletin—sample 30 cents. Dues $7.50/year. Founded 1962.

*IRCA*—International Radio Club of America; P.O. Box 21462, Seattle, WA 98111 . . . Medium-wave radio, DX forum, data on

U.S. AM Stations. Monthly newsletter—sample copy 50 cents. Dues $12.00/year. Founded 1964.

*NRC*—National Radio Club; P.O. Box 127, Boonton, NJ 07005 . . . Pioneer medium-wave club. Broadcast band exclusively. Printed magazine with 30 issues per year—sample copy 50 cents. Dues $15.00/year. Founded 1933.

*NASAWA*—North American Shortwave Association; P.O. Box 13, Liberty, IN 47353 . . . Shortwave bands exclusively. Largest DX club in North America. Monthly bulletin—sample copy $1.00. Dues $13.00/year. Founded 1961.

*SPEEDX;* P.O. Box E, Elsinore, CA 92330 . . . Shortwave, utilities, technical topics. Monthly bulletin—sample copy $1.00. Dues $12.00/year. Founded 1971.

*Worldwide TV-FM DX Association;* P.O. Box 163, Deerfield, IL 60015 . . . The only vhf-uhf DX club. Monthly magazine—free with large SASE. Dues $11.00/year. Founded 1961.

## Associate Members

The following clubs are "associate members." These clubs accept DXers only in the area of their interest.

*Brooklyn DX Club;* 1137 E. 12th Street, Brooklyn NY 11230 . . . Membership limited to those DXers interested in English language broadcasts in shortwave. Monthly bulletin—sample copy 50 cents. Dues $4.50/year. Founded 1975.

*HAP*—Handicapped Aid Program; P.O. Box 163, Mount Sterling, IL 62353 . . . For handicapped DXers and those who are interested in helping the handicapped. Write for information.

*Long Wave Club of America;* 8422 Crane Circle, Huntington Beach, CA 92646 . . . For DXers interested in the frequencies below 550 kHz and in the 1750-meter band. Monthly bulletin—sample copy free with SASE. Dues: 12 large SASEs.

*Miami Valley DX Club;* 1102 Delverne Ave., S.W., Canton, OH 44710 . . . For DXers interested in shortwave band. Monthly bulletin —sample copy 25 cents. Dues $3.00/year. Founded 1973.

*Minnesota DX Club;* 16920–17th Ave., North Wayzata, MN 55391 . . . Limited to DXers living in Minnesota. Monthly bulletin— sample copy SASE. Dues $2.60/year state of Minnesota.

*RCMA*—Radio Communications Monitoring Association; P.O. Box 4563, Anaheim, CA 92803 . . . Emphasis on Southern California. For DXers interested in 30-50, 118-136, 150-174, 225-400 and 450-512 MHz bands. Newsletter. Send SASE for details. Founded 1974.

*Transworld DX Club;* 606 St. Andrews Rd., West Vancouver, British Columbia, Canada V7S 7V4. Limited to DXers living in Canada.

*NNRC*—Newark News Radio Club; P.O. Box 539, Newark, NJ 07101 . . . Shortwave, medium wave, fm, tv, and ham. Monthly bulletin—sample copy $1.00. Dues $13.00/year. Founded 1927.

*EDXC*—European DX Council; c/o Rudolf Heim, P.O. Box 25-03-25, D-4630 Bochum, West Germany. For information on EDXC publications, send four IRCs (International Reply Coupons). The European DX Council has published—in cooperation with Radio Sweden—a list of DX clubs of the world—3 IRCs.

# CHAPTER 6

# Public Service Band Monitoring

Many people are now so familiar with the idea of communications and security that they want to know what is going on. And the public service band monitor radio provides the knowledge of the "action that is going on." Its popularity is tied in with awareness. So, whether you are, besides a DXer, a fire-buff who vicariously "rides the fire truck by radio," or an active photographer hot on the trail of a fast-breaking news story, or a "listener" who likes to keep track of the action around your neighborhood, a public service band monitor is a real must.

Perhaps the largest group of listeners is comprised of those who simply like to listen in for the absolute excitement of it all—to hear the headlines of the next morning while they are taking place—to get an instant look at the behind-the-scenes activities in their community. Maybe it is the rising crime rate, or maybe just that public safety agencies are using their communications facilities more often these days—but today it does seem that it is all more exciting than ever before.

It was not too many years ago that only a few persons listened to the public service bands. There just was not much equipment available which covered the frequencies used by the public service agencies—the vhf and uhf frequencies. As a shortwave DXer, you will note that few shortwave receivers tune above 30 MHz. And the equipment that was on the market tended to be rather expensive. But that has changed. Along with the popularity of listening to the public service bands has come a much wider choice of listening equipment. Prices have dropped, too. Equipment for monitoring the public service frequencies is readily available, even at department stores and discount houses. Probably the most popular are the multiband sets now widely marketed.

Monitoring of the public service bands requires different receiving techniques than those you use for listening to shortwave because the public service bands operate on vhf and uhf frequencies—short range frequencies. There are, basically, three bands to monitor: vhf-low—30 to 50 MHz; vhf-high—150 to 174 MHz, and uhf—450 to 570 MHz.

But monitoring these frequencies raises the issue of the legality of listening to police calls. The popular belief is that it is not only a bit sneaky to do such listening, but also strictly illegal. Unlike shortwave programs, these transmissions are not intended for a general audience. So, can you as a DXer listen in? It is indeed not against any law in the land for you to sit in the privacy of your home and listen to any police, fire, or other transmissions. There is no prohibition *AS LONG AS YOU DO NOT DIVULGE OR USE TO YOUR BENEFIT THE INFORMATION DERIVED FROM YOUR MONITORING.* As an average DXer who monitors just for fun, you are in no danger of being hauled into U.S. Federal Court. In a few areas of the country, there are state or local laws that limit police monitoring, but those restrictions apply *only* to receivers installed in vehicles.

## A CLOSER LOOK AT THE BANDS

There are literally thousands of frequencies—more than 16,000—that you can receive with your monitor. You would obviously want to choose only those frequencies that would be of prime interest to you in your local area to monitor. You may find fire department signals around 33 MHz, 166 MHz, and near 459 MHz (uhf). Some frequencies for police and other law enforcement agencies can be found around 37, 42, 154, 155, 159, 166 MHz (vhf) and 453 and 458 MHz (uhf). But, of course, those signals are not limited to these few areas.

Try the following action-band stations: taxis–152 and 452, trucks–45 and 160, highway-repair crews–37, private police and security agencies–159 and 460, power and light company trucks–46 and 456, and reporters and tv-camera crews–172 to 173 MHz.

Radio-paging stations, which operate a service for those persons who need to keep in touch with their offices or headquarters—even though they are on the go—make use of the higher frequencies. Radio signals alerting clients that messages are waiting are transmitted on 35.22, 35.58, 43.22, and 43.58 MHz. In older types of radio paging, the subscriber hears a beep, which alerts him to the fact that a message is waiting for him. He must then go to a phone and call in to get the message. Other systems operate with voice communications.

Another service you can monitor is the Radio Common Carrier (RCC) mobile telephone. There are 21 channels used by RCC stations across the country.

At something like 60 locations in the U.S., the National Weather Service operates transmitters broadcasting 24 hours a day. The NWS uses three vhf-high band frequencies–162.40, 162.475, and 162.55 MHz.

On the sea and on inland lakes and rivers, there is vhf marine traffic to monitor. Channel 16 on 156.80 MHz is the emergency channel which is monitored by the Coast Guard and which boats are required to monitor as well.

Giant jetliners and private aircraft use the vhf channels between 108 and 136 MHz—on the am frequencies. Because these frequencies cannot be tuned in on most public service band monitors, you will need to invest in another set. There are a number of portable receivers available, however, that will bring you this aero action at bargain prices. Among those frequencies to monitor are the International Distress signal on 121.50 MHz and the Civil Air Patrol emergency frequency on 123.1 and 148.15 MHz. Although vhf is "line-of-sight" you are able to listen to conversations hundreds of miles away because of the broadcaster's plane-mounted antenna often being at 35,000 feet. The airlines party line is on 123.45 MHz. The towers use frequencies between 118 and 120 MHz—the Air Force tower frequency is 126.2 MHz.

There are other signals that can be logged above 30 MHz. These are the two amateur radio bands: the 6- and 2-meter bands. One of the fastest growing areas of ham interest is in the 2-meter fm and repeater operation. With the advent of solid-state design, transceivers can be miniaturized and compact—handheld, portable units are common.

These are just a few examples of the possibilities of vhf/uhf DXing—the field is almost unlimited.

There are three sources where you can obtain the lists of frequencies for your area of interest:

Hollins Radio Data; P.O. Box 35002, Los Angeles, CA 90035.
Monitor Crystal Service; P.O. Box 237, Watseka, IL 60970.
CRB Research; P.O. Box 56, Commack, NY 11725.

Send a stamped, self-addressed envelope to one of these organizations for a price list and catalogue.

# SOME SELECTED FREQUENCIES

Bird's-eye view of the higher frequencies (in MHz).

| | | | |
|---|---|---|---|
| 30–32 | Business | 156–157 | Police/Marine |
| 32–34 | Fire | 158–159 | Police |
| 34–36 | Radio paging | 159–160 | Trucks |
| 36–38 | Highway | 161–162 | Trains |
| | maintenance | 162–163 | Weather |
| 38–40 | Police | 165–167 | Fire |
| 42–46 | Trucks | 169–171 | Fire |
| 46 | Fire | 172–173 | News reporters |
| 46–47 | Power | 452–453 | Taxicabs |
| 47–48 | Special | 453–454 | Police |
| | emergency | 454–455 | Telephone calls |
| 48–50 | Forestry | 456–457 | Power |
| 150–152 | Tow trucks | 458–459 | Police/Fire |
| 152–153 | Taxicabs/Police | 460–461 | Police |
| 154 | Fire | 462–463 | Power |
| 154–155 | Police | 465–466 | Police/Fire |
| 155–156 | Special | | |
| | Emergency | | |

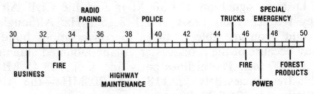

*(A) The am band—30 to 50 MHz.*

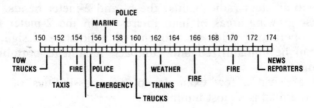

*(B) The vhf-fm band—150 to 174 MHz.*

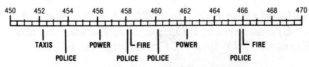

*(C) The uhf-fm band—450 to 470 MHz.*

**Fig. 6-1. The fm public service bands.**

Fig. 6-1 provides you with a graphic display of the fm public service bands. The public service bands cover ten thousand different frequencies—too many to list here. However, in the pages following there are some samples of a number of interesting frequencies.

The tv broadcast stations operate on the following frequencies:

| 55.25– 87.75 | Channels | 2– 6 |
| 175.25–215.75 | Channels | 7–13 |
| 471.25–889.75 | Channels | 14–83 |

## RADIOTELEPHONE FREQUENCIES (RCC)

| vhf-high band | Receive | Transmit | Channel |
|---|---|---|---|
| | 152.03 | 158.49 | 1 |
| | 152.06 | 158.52 | 3 |
| | 152.09 | 158.55 | 5 |
| | 152.12 | 158.58 | 7 |
| | 152.15 | 158.61 | 9 |
| | 152.18 | 158.64 | 11 |
| | 152.21 | 158.67 | 13 |
| uhf band | 454.025 | 459.025 | 21 |
| | 454.050 | 459.050 | 22 |
| | 454.075 | 459.075 | 23 |
| | 454.100 | 459.100 | 24 |
| | 454.125 | 459.125 | 25 |
| | 454.150 | 459.150 | 26 |
| | 454.175 | 459.175 | 27 |
| | 454.200 | 459.200 | 28 |
| | 454.225 | 459.225 | 29 |
| | 454.250 | 459.250 | 30 |
| | 454.275 | 459.275 | 31 |
| | 454.300 | 459.300 | 32 |
| | 454.325 | 459.325 | 33 |
| | 454.350 | 459.350 | 34 |

The following is a selected list of frequencies used by police departments in major cities.

| City | Frequencies |
|---|---|
| Akron, OH | 155.97/156.15/156.21/460.050/460.375 |
| Atlanta, GA | 155.85/156.21/460.025/460.075/460.150/460.425 |
| Baltimore, MD | 155.79/495.1875 |
| Cleveland, OH | 460.125/460.225/460.275/460.400/460.450 |
| Columbus, OH | 154.65/154.71/155.25/155.55/155.58/460.075/ 460.125/460.200/460.275/460.350/460.400 |
| Dallas, TX | 155.19/155.26/460.025/460.075/460.175 |
| Denver, CO | 156.03/460.425 |
| Detroit, MI | 155.85/460.275 |
| Houston, TX | 154.86/155.13/155.55/460.025/460.050/460.350 |

| City | Frequencies |
|------|-------------|
| Kansas City, MO | 154.71/155.64/155.85/156.09/453.10 |
| Las Vegas, NV | 155.55/155.73 |
| Little Rock, AR | 155.61 |
| Los Angeles, CA | 154.83/159.03/453.100/506.5875/506.8125 |
| Minneapolis, MN | 154.74/154.80/156.03 |
| New Orleans, LA | 158.91/460.25/460.100/460.200/460.225/460.475 |
| New York, NY | 470.6375/470.6625/470.6875/470.7125/470.7375/ 470.8125/470.8625/470.8875/470.9375/ 470.9875 |
| Omaha, NB | 460.100/460.225/460.375 |
| Philadelphia, PA | 155.07/155.625/460.150/460.250/460.350/460.375 |
| Pittsburgh, PA | 158.97/**458.10/458.40** |
| Portland, OR | 154.65/155.97/156.15/158.91/159.21/460.500 |
| St. Louis, MO | 155.61/159.03/453.40 |
| Salt Lake City, UT | 154.83/154.875/155.01/155.13 |
| San Francisco, CA | 45.10/45.14/45.46/45.58/460.125/460.350/460.075 |
| Seattle, WA | 155.25/155.64/155.97 |
| Washington, DC | 460.100/460.250/460.475 |

The following is a selected list of frequencies used by fire departments in major cities.

| City | Frequencies |
|------|-------------|
| Atlanta, GA | 153.89/154.19/460.600/460.625 |
| Baltimore, MD | 154.07/154.31 |
| Cheyenne, WY | 154.355 |
| Cleveland, OH | 153.89/153.93/153.95 |
| Chicago, IL | 153.83/153.95/154.01/154.13 |
| Columbus, OH | 153.77/154.31 |
| Denver, CO | 153.77/154.31 |
| Dallas, TX | 153.89/154.07/154.22 |
| Houston, TX | 153.89/154.16/154.28 |
| Kansas City, MO | 154.01/154.13/154.28/458.10/462.95 |
| Little Rock, AR | 153.95 |
| Los Angeles, CA | 33.48/33.53/506.3125/506.6375/507.0125 |
| Las Vegas, NV | 154.37 |
| New Orleans, LA | 154.235/154.37/453.15/460.575/460.600/460.625 |
| Omaha, NB | 154.01/154.19 |
| Philadelphia, PA | 153.89/153.95/154.01/154.145/154.23 |
| Portland, OR | 154.25/154.28/460.525/460.550/460.575 |
| Pittsburgh, PA | 154.13/154.445 |
| New York, NY | 153.77/153.89/153.95/154.01/154.07/154.19/ 460.525/460.575/460.625/460.675 |
| Salt Lake City, UT | 154.31/154.34/154.43 |
| San Francisco, CA | 154.28/154.43 |
| Seattle, WA | 154.19/154.25/154.34/453.70/453.80 |
| Washington, DC | 154.19/154.205/154.235 |

Now that you have an idea on which frequencies to listen for the police and fire departments, you will hear conversations in a number of different codes—police codes. On the following pages, a number of these codes are listed as they may be used by law enforcement agencies.

## MISCELLANEOUS CODES

The following is a 10-code chart most often used by law enforcement agencies.

| | | | |
|---|---|---|---|
| 10–0 | Caution | 10–37 | Investigate suspicious vehicle |
| 10–1 | Unable copy—change location | 10–38 | Stopping suspicious vehicle |
| 10–2 | Signal good | 10–39 | Urgent, use light, siren |
| 10–3 | Stop transmitting | 10–40 | Silent run—no light, siren |
| 10–4 | Acknowledgment—OK | | |
| 10–5 | Relay | 10–41 | Beginning tour of duty |
| 10–6 | Busy—unless urgent | 10–42 | Ending tour of duty |
| 10–7 | Out of service | 10–43 | Information |
| 10–8 | In service | 10–44 | Permission to leave for . . . |
| 10–9 | Repeat | | |
| 10–10 | Fight in progress | 10–45 | Animal carcass at . . . |
| 10–11 | Dog case | 10–46 | Assist motorist |
| 10–12 | Standby—stop | 10–47 | Emergency road repair at . . . |
| 10–13 | Weather—road report | | |
| 10–14 | Prowler report | 10–48 | Traffic standard repair at . . . |
| 10–15 | Civil disturbance | | |
| 10–16 | Domestic problem | 10–49 | Traffic light out at . . . |
| 10–17 | Meet complainant | 10–50 | Accident (F, PI, PD) |
| 10–18 | Quickly | 10–51 | Wrecker needed at . . . |
| 10–19 | Return to | 10–52 | Ambulance needed at . . . |
| 10–20 | Location | 10–53 | Road blocked at . . . |
| 10–21˙ | Call . . . by phone | 10–54 | Livestock on highway |
| 10–22 | Disregard | 10–55 | Intoxicated driver |
| 10–23 | Arrived at scene | 10–56 | Intoxicated pedestrian |
| 10–24 | Assignment completed | 10–57 | Hit and run (F, PI, PD) |
| 10–25 | Report in person (meet) | 10–58 | Direct traffic |
| 10–26 | Detaining subject— expedite | 10–59 | Convoy or escort |
| | | 10–60 | Squad in vicinity |
| 10–27 | Drivers license info. | 10–61 | Personnel in area |
| 10–28 | Vehicle registration information | 10–62 | Reply to message |
| | | 10–63 | Prepare make written copy |
| 10–29 | Check for wanted | | |
| 10–30 | Unnecessary use of radio | 10–64 | Message for local delivery |
| 10–31 | Crime in progress | | |
| 10–32 | Man with gun | 10–65 | Net message assignment |
| 10–33 | Emergency | 10–66 | Message cancellation |
| 10–34 | Riot | 10–67 | Clear for net message |
| 10–35 | Major crime alert | 10–68 | Dispatch information |
| 10–36 | Correct time | | |

| | | | |
|---|---|---|---|
| 10–69 | Message received | 10–87 | Pickup/distribute checks |
| 10–70 | Fire alarm | 10–88 | Present phone number |
| 10–71 | Advise nature of fire | | of . . . |
| 10–72 | Report progress on fire | 10–89 | Bomb threat |
| 10–73 | Smoke report | 10–90 | Bank alarm at . . . |
| 10–74 | Negative | 10–91 | Pickup prisoner/ |
| 10–75 | In contact with | | subject at . . . |
| 10–76 | En route | 10–92 | Improperly parked |
| 10–77 | ETA | | vehicle |
| 10–78 | Need assistance | 10–93 | Blockade |
| 10–79 | Notify coroner | 10–94 | Drag racing |
| 10–80 | Chase in progress | 10–95 | Prisoner/subject in |
| 10–81 | Breathalyzer report | | custody |
| 10–82 | Reserve lodging | 10–96 | Mental subject |
| 10–83 | Work school xing at | 10–97 | Check (test) signal |
| 10–84 | If meeting . . . advise | 10–98 | Prison/jail break |
| 10–85 | Delayed due to . . . | 10–99 | Wanted/stolen indicated |
| 10–86 | Officer/operator on duty | | |

The following is the 11-code, also sometimes used by law enforcement agencies.

| | | | |
|---|---|---|---|
| 11–6 | Illegal discharge of firearms | 11–43 | Doctor required |
| 11–7 | Prowler | 11–44 | Coroner required |
| 11–8 | Person down | 11–45 | Attempted suicide |
| 11–10 | Take a report | 11–46 | Death report |
| 11–12 | Dead animal | 11–47 | Injured person |
| 11–13 | Injured animal | 11–48 | Provide transportation |
| 11–14 | Animal bite | 11–65 | Traffic signal light out |
| 11–15 | Ball game in street | 11–66 | Traffic signal out of order |
| 11–17 | Wires down | 11–70 | Fire alarm |
| 11–24 | Abandoned vehicle | 11–71 | Fire report |
| 11–25 | Vehicle—traffic hazard | 11–79 | Traffic accident— ambulance sent |
| 11–25X | Female motorist needs assistance | 11–80 | Traffic accident— serious injury |
| 11–27 | Subject has felony record but is not wanted | 11–81 | Traffic accident— minor injury |
| 11–28 | Rush vehicle registration information—driver is being detained | 11–82 | Traffic accident— no injury |
| 11–29 | Subject has no record, and is not wanted | 11–83 | Traffic accident— no details |
| 11–30 | Incomplete phone call | 11–84 | Direct traffic |
| 11–31 | Person calling for help | 11–85 | Dispatch towtruck |
| 11–40 | Advise if ambulance is needed | 11–86 | Special detail |
| 11–41 | Request ambulance | 11–87 | Assist other unit |
| 11–42 | Ambulance not required | 11–98 | Meet officer |
| | | 11–99 | Officer needs help |

Yet, another code used by some law enforcement agencies is the 900-code.

| | | | |
|---|---|---|---|
| 901 | Ambulance call | 920A | Found adult |
| 902 | Accident | 920C | Missing child |
| 903 | Aircraft crash | 920F | Found child |
| 904 | Fire | 921 | Prowler |
| 905B | Animal bite | 921P | Pepping tom |
| 905V | Vicious animal | 922 | Illegal peddling |
| 906 | Officer needs minor assist | 924 | Station detail |
| 906C | Officer needs assist with vehicle | 925 | Suspicious person |
| | | 926 | Request towtruck |
| 907 | Minor disturbance | 926A | Towtruck dispatched |
| 907A | Loud radio or tv | 927 | Investigate unknown trouble |
| 907B | Ball game in street | | |
| 908 | Begging | 927A | Person pulled from phone |
| 909 | Traffic congestion | | |
| 909B | Road blockade | 927D | Investigate possible dead body |
| 909F | Flares needed | | |
| 909T | Traffic hazard | 928 | Found property |
| 910 | Can you handle call? | 929 | Investigate person down |
| 911 | Advise party | 930 | See man regarding complaint |
| 911B | Contact informant | | |
| 912 | Are we clear? | 931 | See woman regarding complaint |
| 913 | You are clear | | |
| 914 | Request detectives | 932 | Woman or child being abused |
| 914D | Request doctor | | |
| 915 | Dumping rubbish | 981 | The frequency is clear |
| 916 | Holding suspect | 982 | Are we being received? |
| 917A | Abandoned vehicle | 995 | Strike trouble |
| 917P | Hold vehicle for fingerprints | 996 | Explosion |
| | | 996A | Unexploded bomb |
| 918 | Mental case | 998 | Officer involved in shooting |
| 918A | Escaped mental patient | | |
| 918V | Violent mental case | 999 | Officer needs help— urgent |
| 919 | Keep the peace | | |
| 920 | Missing adult | | |

Often the DXer will hear the following codes used:

| | | |
|---|---|---|
| Code | 1 | Acknowledge this call |
| Code | 2 | Proceed immediately without siren |
| Code | 3 | Proceed immediately with siren |
| Code | 4 | No further assistance required |
| Code | 4A | No further assistance required—suspect is in custody |
| Code | 5 | Stakeout—uniformed officers stay away from location |
| Code | 6 | Out of vehicle for investigation |
| Code | 6A | Out of vehicle for investigation—assistance may be required |
| Code | 6C | Subject wanted—may be dangerous |
| Code | 7 | Out of service to eat |

| Code | 8 | Fire alarm |
|------|-----|------------|
| Code | 12 | Patrol your district and report extent of disaster damage |
| Code | 13 | Major disaster activation |
| Code | 14 | Resume normal operation |
| Code | 20 | Notify newsmedia to respond |
| Code | 33 | Emergency traffic on the air—all units stand by |

# CHAPTER 7

# Selecting Your Equipment

As described earlier, the vhf and uhf bands are further up the radio-frequency spectrum from the short and medium waves and unless you have one of the multiband radios, you will need a second receiver to tune to these higher frequencies.

The moment will come for you to decide which type of public service band receiver you want to use. But first take stock of your needs. Which agencies do you want to monitor? Do you monitor in your car or at home? Two simple, basic questions—but only after you have properly answered them, can you evaluate the equipment and features you want to purchase. So, let's see what the market basically has to offer.

## CHOOSING A RECEIVER

Monitoring equipment for picking up the public service bands comes in a variety of packages and styles. For the casual listener, a multiband receiver with coverage of both low- and high-band vhf provides plenty of listening activity. If you have a special interest be sure the frequency coverage of the receiver includes those bands. Check the manufacturer's literature to determine exactly what frequencies you will receive. These multiband sets are good choices for casual or occasional listening because they cover large segments of the monitor spectrum at relatively low cost. Tuning is like that of an ordinary table radio; these receivers are called "continuous-tuning receivers." A continuously tunable receiver has two inherent disadvantages: *inaccurate frequency calibration* and *drift*. The dials of most receivers are only marked every 3 or 4 MHz, so you can only estimate the position of the desired channel on the dial. Drift occurs when the receiver is turned on after it has been switched off for a few hours and you find a different station on the same position that you had before.

## CRYSTAL-CONTROLLED RECEIVER

A step-up in receiver specialization is the monitor-only receiver. This type of receiver limits its coverage to one or more of the vhf or uhf bands. Although higher in price than the continuous-tuning receiver, these sets offer certain advantages.

Most receivers, including crystal-controlled receivers, have a squelch control that allows the speaker to remain silent until an incoming signal unlocks the squelch. The basic advantage of this type of unit is, of course, its crystal control that provides greater ease and accuracy of tuning. Although the cost of each crystal carries a price tag of about $5.00, you can lock the receiver to the precise frequency desired without drift.

## SCANNER-MONITOR

The *scanner* is the most sophisticated, specialized receiver. It solves two of the problems common when you have another kind of receiver. First, a lot of time and dead air may pass between transmissions while you listen and wait for some action. Second, when you leave your receiver on one frequency, you may completely miss the action on another. Thus, the scanner provides a neat electronic solution with no moving parts—it silently searches over a number of frequencies and stops when it senses an incoming carrier. Audio is heard in the speaker as long as the carrier is present. When the transmission ends, the scanner resumes its exploration over the band for more signals. So, with no effort on your part, a scanner vastly increases listening opportunities and reduces idle periods.

The first receiver sets were typically single band, 8-channel scanners. The multiband scanner (like those shown in Figs. 7-1 and 7-2) was introduced to permit the DXer to hear channels scanned in more than one band. The GR-1132 model in Fig. 7-1 is a crystal-controlled, 3-band, 8-channel scanner capable of covering any 8-MHz segment of vhf-low (30 to 50 MHz), vhf-high (148 to 174 MHz), and uhf-low (450 to 470 MHz). The front-mounted controls include: on/off/volume control, squelch control, scan select button, and automatic/manual scan select. It features built-in priority channel override. This scanner has three telescoping antennas and operates on 12 vdc or 117 vac. Fig. 7-2 shows another crystal-controlled scanner, Model 4354, that features individual lock-out

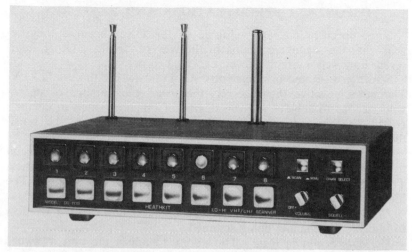

Fig. 7-1. Model GR-1132 crystal-controlled, 3-band, 8-channel scanner.

Fig. 7-2. Model 4354 crystal-controlled, 12-band scanner.

switches. This 12-band scanner covers vhf-low (30 to 50 MHz), vhf-high (148 to 174 MHz), uhf-low (450 to 470 MHz), and uhf-high (470 to 512 MHz). Front controls include: on/off/volume control, squelch control, priority switch, scan delay, and weather monitoring button. It has a built-in vhf/uhf antenna and operates on 12 vdc or 117 vac.

## IMPORTANT SPECIFICATIONS AND FEATURES

Some specifications are valuable to the DXer because they reveal what the scanner contains in the way of circuitry. Other items reveal how well a circuit can be expected to perform.

*Sensitivity*—Rated in microvolts and indicated by the symbol $\mu V$. The more sensitive the receiver, the fewer the number of microvolts needed to achieve a given degree of quieting; rated in decibels (dB). The lower the number of microvolts, the better the sensitivity.

*Selectivity*—Rated in decibels (dB). It is the ability of a circuit to reject signals which are just above or below the desired frequency. The higher the selectivity, the more it eliminates interference. The higher the decibel rating is, the better the selectivity.

*Digital Readout*—A computer-like numeral flashes in a window to indicate the channel being scanned.

*Variable Scanning Rate*—The ability to adjust the unit to sweep the band or bands from 2 to 50 channels per second.

*Priority Channel*—Allows normal scanning to be interrupted whenever a signal is received on a preselected or priority frequency.

## SYNTHESIZED SCANNERS

A synthesized receiver such as the one in Fig. 7-3 does not require a crystal for each channel. Instead, each position in the scan-

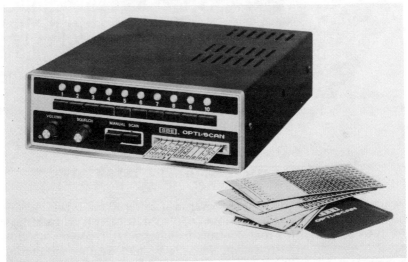

Courtesy SBE Inc.

**Fig. 7-3. Opti-Scan (SBE-12SM) scanning monitor-receiver.**

ner can be programmed in various ways to receive any frequency within the range of the receiver capability. Otherwise, the scanner functions in the same manner as described previously. The Opti/Scan Model 12SM in Fig. 7-3 is a synthesized all-band, 10-channel scanner. It covers all frequency bands from 30 to 512 MHz. No crystals are required—optical cards are used to pick up 16,000 frequencies. Each card may contain ten frequencies. Front controls include: volume control, squelch control, automatic/manual scan control, and individual lock-out switches. The Model 12SM operates on 12 vdc or 117 vac.

## COMPUTERIZED SCANNERS

New technological advances nearly always require the sacrifice of features previously available, . . . like exact frequency control by means of a crystal.

Now there are several scanners available like the Bearcat 210 shown in Fig. 7-4 that can search the entire vhf and uhf bands to detect any active frequencies and store them in a random-access memory. Later these frequencies can be recalled automatically, or manually by pressing a button.

Courtesy Electra Co.

**Fig. 7-4. Bearcat 210 computerized scanner.**

The Bearcat 210 computerized scanner features a microprocessor and provides unlimited search capability. A push-button keyboard allows access to more than 6,000 public service frequencies within the 10 channels available. Scan rate is 20 channels per second. This scanner covers all bands from vhf-low to uhf-high. Other features include: decimal LED frequency display, selective scan delay, automatic lock-out control, and automatic frequency search. Front controls include: on/off/volume control, scan control, and manual control. It has a telescoping antenna and operates on 12 vdc or 117 vac.

The heart of the system in these scanners is a microprocessor, which provides control of frequencies through a phase-locked loop frequency synthesizer that eliminates the need for channel crystals. The scanners also differ from the synthesized scanners in the following manner: . . . the synthesized scanners described above require the use of a code book to set up new frequencies, but the computerized scanners use a keyboard for each frequency search. The frequency selected and being received by the unit is displayed by means of an LED numerical display.

## POCKET SCANNERS

There is a category of receivers available to the public service band DXer that is not available to the shortwave DXer: the pocket scanner. The PSB enthusiasts can take their monitoring capabilities with them everywhere they go. The modern pocket scanner fits on your belt or in your pocket. Scanners come in several variations—for low band only, high band only, or a combination of both. There is no triband pocket scanner available at this writing, however.

Fig. 7-5 shows a handheld, crystal controlled, all-band, 4-channel pocket scanner. It covers all bands from 30 to 512 MHz and has individual lock-out switches and telescoping antenna. Controls include: on/off/volume control, squelch control, and automatic/manual channel select. This scanner is battery powered by four "AA" cells.

The pocket scanners have the features the mobile/base units have. They have a volume and squelch control. Many of them have a scan/manual switch, allowing you to scan all four channels (most pocket scanners have four channels) or to hold any one of the four you select. Models with individual lock-out switches and LED channel readout as shown in Fig. 7-6 also are available.

The handheld scanner shown is a Pockette crystal-controlled, 2-band, 4-channel model that covers uhf-low and high. It has on/off/volume control, squelch control, automatic/manual channel-select switch, and individual lock-out switches. The Pockette model is equipped with telescoping antenna, external-antenna jack and operates on four "AA" batteries.

**Fig. 7-5. Bearcat handheld, crystal-controlled, all-band, 4-channel pocket scanner.**

All pocket scanners operate on batteries—either regular cells or Ni-Cad (nickel-cadmium—rechargeable) batteries. These scanners operate quite efficiently on their built-in telescoping or rubber-covered antenna.

Sensitivity of the pocket scanner easily matches that of the mobile or base scanner. They lack the sensitivity of the large scanners because of the tightness of their circuitry.

## ANTENNAS

Because of the way signals travel at most communication frequencies, DXers have found it advisable to mount the receiving antenna as high as possible for maximum practical coverage. Mounting

Courtesy RCA Corp.

**Fig. 7-6. Pockette model crystal-controlled, 2-band, 4-channel handheld scanner.**

a fixed antenna on the roof has extended coverage many miles past where the built-in antenna of the receiver would have failed. Yet, the whip antenna attached to most of the scanners works well if you wish to monitor local signals only.

For use with a single-band or multiband scanner, you have a choice of wideband antennas for both mobile and home applications. There are multiband groundplane antennas as well as antennas that can be used at fixed locations for receiving on all four of the public service bands.

A single-band receiver has only one antenna jack, usually one that mates with a Motorola-type auto-radio antenna plug. A low-high vhf scanner usually has one similar antenna jack through which signals from both bands are fed. A multiband receiver usually has separate antenna jacks for vhf and uhf inputs.

If you want more coverage you can use an antenna with "gain." Gain is the ability of the antenna to multiply the strength of the in-coming signal—it is measured in dB. The higher the dB rating, the better is the ability of the antenna to pull in signals. An antenna offer-ing 3-dB gain will double the strength of the incoming signal.

Another factor is directivity—the ability of the antenna to hear best in only one direction. Such antennas are called "beam" or "yagi" antennas and the decibel ratings apply only to the one direction that they will pick up. An antenna, however, that receives with equal ability in all directions is called *omnidirectional*. This kind is the best for general use.

## Mobile Antennas

If you install a scanner in your car, your range depends greatly on the effectiveness of the antenna. For a single-band receiver, you can use a quarter-wave whip—6 feet long for 30 to 50 MHz; 18 inches long for 150 to 174 MHz; 6 inches long for 450 to 470 MHz and 5 inches long for 470 to 512 MHz. Preferably, it should be mounted in the center of the metal car roof for proper ground plane effect. A typical example of a commercial mobile antenna is shown in Fig. 7-7.

Unless you have a very high antenna, do not count on hearing mobile units from anywhere nearly as far as base stations with tall antennas and, generally, more powerful transmitters.

There are different styles of mobile antennas available. Some whips 6 feet long are available with ball-type mounting for side-mount installations. The same length whip also comes in a bumper-mount style. The smaller quarter-whip antennas are available in roof-top mount (requiring that a hole be drilled through the vehicle roof top), gutter-mount, magnetic-mount, and trunklid-mount configura-tions.

## Mobile Receiver and Antenna Installation

There are a number of possible mounting locations for the mo-bile receiver. One is under the dashboard. Another is the console mount or floor mount. Whatever location is the most convenient for your vehicle, always bear in mind the following:

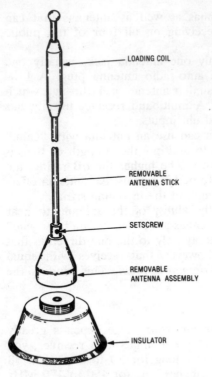

LOADING COIL

REMOVABLE
ANTENNA STICK

SETSCREW

REMOVABLE
ANTENNA ASSEMBLY

INSULATOR

**Fig. 7-7. Typical commercial mobile-station antenna.**

(1) The controls of your scanner must be convenient and visible.
(2) The location of your scanner should not interfere with your normal functions as a driver.
(3) The scanner should not be mounted in the way of heater ducts, air-conditioning outlets, or direct air blast inlets.
(4) The scanner should be protected from rain or spray.

With the aid of some basic tools and an electric drill, installation of a scanner is not a major task. Secure the mounting bracket and attach the scanner to it.

When connecting the scanner to a power source, take the following steps:

(1) Determine the polarity of the scanner. The red wire coming from the set—usually containing a fuse—is the positive, hot wire. The black wire is negative, ground.
(2) Find a convenient power source. The best one is the positive side of the battery. An alternative is the fuse block, located under the dash.
(3) Use a voltmeter to find which vacant terminal posts on the fuse block are "live."

(4) Connect the red wire to the block.

(5) Connect the ground lead to a convenient bolt or screw. It is important to make solid connections or you will experience noise interference.

For easy scanner removal, an alternate connection can be made by using the cigarette lighter socket.

Once you have selected the type of antenna to use and chosen the location on your vehicle, study the manufacturer's mounting instructions. Should you decide on a trunklid-mount antenna, even the routing of the antenna cable is not too complicated.

(1) To find a passage between the trunk and the passenger compartment, insert a stiff wire through one of the lower corners of the trunk. Most of the time you will find a small opening under the rear seat.

(2) Pull the antenna cable through.

(3) Loosen the metal molding along the sides of the car floor. This allows the cable to be stuffed under the mat or carpet and routed up behind the dashboard for connection.

Always try to keep the cable away from areas where it would be subject to excessive abrasion or strain.

### Base Antennas

Should you decide against using the telescoping whip supplied by the manufacturer, a popular stationary antenna to use is the ground-plane antenna like that shown in Fig. 7-8. This antenna consists of a vertical radiator (or whip) with three or four radial elements protruding from its base. These radials create the electrical ground

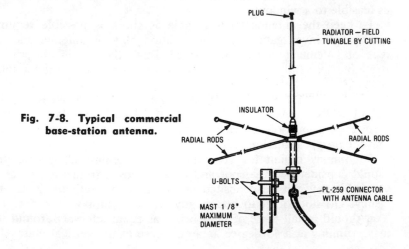

**Fig. 7-8. Typical commercial base-station antenna.**

plane required by the vertical whip. The ground plane antenna may be located at any height above the ground. This antenna is also omni-directional.

There are several types of antennas available—for low-band frequencies only or for high-band frequencies. There are also dual-band antennas on the market. Be sure to get the right antenna for the type of receiver you use.

At fixed locations, antenna transmission-line losses can become significant. It is desirable to limit your signal loss. For example, to obtain 3 dB (10% signal loss) when using RG-58 coaxial cable, you can use 100 feet at 50 MHz, 50 feet at 160 MHz, and 25 feet at 450 MHz. So, if you need to use 50 feet of cable to get from your antenna to your scanner, RG-58 cable will be all right when you want to monitor the low frequencies, i.e., the vhf band. Should you want to monitor the uhf band in this situation, you would lose about 6.5 dB, or 53% of the signal. If you use the more expensive RG-8 coaxial cable, your loss will again be in the 3-dB range.

Increasing antenna elevation 3.75 times achieves a 10-dB gain, or 320% signal voltage increase.

### Base Receiver and Antenna Installation

Some consider the installation of a base-station scanner easier than a mobile rig. Others claim that a base station is more difficult to install. It all depends what YOU want. Where you plan to install your scanner . . . the type of antenna used . . . type of transmission line and its length . . . type of structure you live in . . . your neighbor's tv antenna location . . . your own tv antenna location . . . etc.

The scanner itself should be placed as far away as possible from the tv set, as close as practical to a 117-volt ac outlet, and as near as feasible to a window.

To keep the antenna lead-in cable as short as possible requires a thorough investigation into the location of your antenna and the type of antenna you have purchased. Read the manufacturer's instructions carefully. Check the tools you need. Use rubber-soled shoes if you climb on the roof—this prevents you from slipping. CONSIDER THE ELECTRICAL WIRES AND POWER LINES THAT RUN CLOSE TO YOUR ROOF!! Use a good ladder. Trees close to the location where you want to mount your antenna will prevent proper radiation.

A chimney mount for your antenna is the last alternative you should consider because soot and other deposits from the smoke of your chimney can be deposited on the antenna and lower its efficiency. Do not forget to attach guy wires to support the antenna. You should install three of these wires at equal intervals around the mast, running at a 45-degree angle or more to the vertical mast.

After installing the antenna you must rout the lead-in cable. How you run the cable depends on the type of house you have and the location of your scanner. You might drill a hole in a wall and run the cable upward alongside the outside wall. Another way might be to go straight up along the inside wall to the attic and through a roof vent to the outside. You could also drill a hole in the floor and run the cable through the crawl space below the house and via a bottom vent along the outside wall to the roof. Caulk each hole that you must drill, or outside weather will get inside the house. Once you have reached the roof, you can let the cable lie flat on your rooftop or use standoff insulators.

After you have completed the installation and before you plug in the cable to the scanner, install a lightning arrestor. Attach this arrestor to the antenna connector and run a 12- or 14-gauge copper wire from the arrestor to a cold-water pipe or other ground rod (See Chapter 4.) This will ground your system properly against lightning strikes.

## IN A NUTSHELL

This is DXing in a nutshell! Curious? Excited? Go out, buy your DX receiver and scanner, and find out for yourself. I am sure that you will be "bitten by the DX-bug," just as I was when I lived and worked in the swamps and jungles of Borneo. There was one thing that kept my spirits high and relieved the boredom of those lonely tropical nights: my DX receiver powered by the battery of my jeep. I was no longer isolated. . . . I could hear "the world." I knew that another world existed besides trees . . . and foliage . . . and water . . . and mosquitos. That there were sounds of music . . . and voices (other than the croaking of the frogs . . . the chattering of the monkeys . . . and all the other jungle sounds).

From that time on, I have always had a shortwave receiver. My writing prevents me from regularly scanning the bands. But when I tire of pounding the typewriter, I just turn on my receiver and the sounds of the world enter my ears—I am a DXer again. So, try it.

Happy hunting!! And you know, you do not need a "hunting license." What are your "hunting grounds"? Well, there are a multitude of wave bands to listen to in your neighborhood and around the world.

# Glossary

## A

*adjacent channel interference:* Interference from a station whose frequency is slightly below or above that of the desired station.

*alligator clip:* A long-nosed metal clip with meshing jaws, generally used to make temporary connections.

*all-wave antenna:* An antenna designed to receive or radiate a wide range of radio frequencies.

*all-wave receiver:* A radio receiver capable of broadcast and shortwave reception. A common all-wave receiver tunes from about 500 kHz to 30 MHz.

*alternating current:* A term used to distinguish current of changing polarity from direct or constant polarity.

*amateur:* The group of validly licensed, noncommercial radio operators, usually referred to as "hams."

*amateur bands:* Radio-frequency bands assigned to radio amateurs by international agreement.

*amplitude modulation:* The modulating of a carrier-frequency current by varying its amplitude above and below normal value in accordance with the intelligence being transmitted.

*antenna:* A system of electrical conductors used for reception of radio waves. Specifically: a radiator for the reception of electromagnetic radio waves. A synonym for aerial.

*antenna coil:* The rf coil in a radio receiver to which the antenna is connected.

*antenna coupler:* A device used to connect a receiver to an antenna or antenna line.

*antenna current:* The current flowing in the antenna and associated circuits.

*antenna farm:* Refers to the collection of antennas used by shortwave broadcasters for beaming their signals throughout the world.

*antenna frequency:* The highest frequency that is returned to earth during a given transmission—also known as the "critical" antenna frequency.

*atmospheric interference:* Crackling and hissing noises reproduced by a receiver as a result of electric disturbances in the atmosphere. Also called "static."

*attenuation:* Reduction in the strength of an electric impulse.

*attenuator:* A fixed or variable device used to reduce the amplitude of an electric impulse.

*audio:* Voltages or current in the audible-frequency range.

*audio-frequency:* A frequency in the range of audible sound waves. The audio-frequency spectrum is from 15-20,000 cps (cycles per second).

*automatic frequency control:* A circuit which keeps a receiver accurately tuned to a predetermined frequency.

*automatic volume control:* A circuit which automatically maintains a constant output volume in spite of varying input signal. Used in practically all modern receivers where it minimizes fading and prevents blasting when tuning suddenly from a weak station to a strong one.

*aurora zone:* An ionized polar sky cap—over both North and South poles—Auroras make reception over the poles very difficult most of the time. Sometimes they act as "mirrors" and bring in unique DX signals.

# B

*band:* Those frequencies that are within two definite limits. For example, the standard broadcast band extends from 550 to 1600 kHz.

*bandspread:* Any method, mechanical or electronic, of effectively increasing the tuning scale of a receiver between radio stations.

*band switch:* A switch used to change one or more circuits of a multi-band radio receiver from one band to another; also called band selector.

*bandwidth:* A section of the frequency spectrum required to transmit the desired intelligence. For example, the bandwidth of the average am broadcast channel is 10 kHz.

*beam antenna:* Antenna array that receives radio frequencies more sharply in one direction than the others.

*beat-frequency oscillator:* A device from which an audible signal is obtained by combining and rectifying two higher inaudible frequencies.

*broadcast band:* The band of frequencies between 550 and 1650 kHz in which are assigned all standard am broadcast stations operating in the U.S.

*built-in antenna:* An aerial which is an integral part of a receiver, such as a compact loop aerial.

# C'

*call letters:* Assigned identifying letters for a radio station. Letters are assigned by the FCC and by authorized branches of the government.

*carrier:* A current, voltage, or radio wave at the assigned frequency of a radio station.

*cell:* A dc voltage source. A dry cell cannot be recharged when exhausted. A storage cell can be recharged when exhausted by passing a current through it.

*center frequency:* The assigned frequency of an fm station. Frequency shifts take place in step with the audio signal.

*channel:* (1) A band of frequencies including the assigned carrier frequency, within which transmission is confined in order to prevent interference with stations on adjacent channels. (2) An electrical path over which signals travel, thus, an amplifier may have several input channels, such as microphone, tuner, or phonograph.

*coaxial cable:* Also coax cable or coax. A two-conductor cable in which one conductor is a flexible or nonflexible metal tube and the other is a wire axially supported inside the tube by insulators.

*code:* A system of dot and dash signals used in the transmission of messages by radio or wire telegraphy. The International Morse code (also called the Continental code) is used universally for radiotelegraphy. The American Morse code is used commonly for wire telegraphy.

*continuous wave:* An unmodulated, constant-amplitude radio-frequency wave.

*crystal:* A piece of quartz or similar piezoelectric material which has been ground to proper size to produce natural vibrations at a desired radio frequency. Quartz crystals are used in radio receivers to generate with a high degree of accuracy the assigned carrier frequency.

# D

*decibel (dB):* A term expressing a ratio between two amplitudes or energies. The decibel unit between two amplitudes is computed as twenty times the log of the ratio—between two energies as ten times the log of the ratio. Practically, a decibel is approximately the smallest change in sound intensity that the human ear can detect.

*dipole antenna:* A conductor one-half wavelength long at a given frequency. Used to pick up radio waves.

*direct current (DC):* An electric current which flows in only one direction.

*directional antenna:* Any antenna which picks up or radiates signals better in one direction than another.

*distress frequency:* The frequency allotted to distress calls by international agreement.

*distress signal:* By code S.O.S. By radiotelephone "mayday."

*DX:* Code or radiotelephonic colloquialism for the word "distance"—referring to distant reception of radio signals.

# E

*E-layer:* An ionized layer in the E-region of the ionosphere.

*electromagnetic spectrum:* Range of frequencies of electromagnetic waves.

| Type of Wave | Wavelength |
| --- | --- |
| radio | above 1000 km to below 1 cm |
| infrared | 0.03 to 0.000076 cm |
| visible light | 0.000076 to 0.000040 cm |

| ultraviolet | 0.000040 to 0.0000013 cm |
| X-rays | $10^{-6}$ to $10^{-9}$ cm |
| gamma rays | $10^{-8}$ to $5 \times 10^{-11}$ cm. |
| cosmic rays | $10^{-11}$ to $10^{-12}$ cm |

*E-region:* A region in the ionosphere, from about 55 to 85 miles above the surface of the earth, containing ionized layers capable of bending or reflecting radio waves.

## F

*fading:* A variation in the signal intensity at a given location or at a given dial setting of a receiver.

*F-layer:* An ionized layer in the F-region of the ionosphere. About 90 to 250 miles above the surface of the earth.

*frequency:* The number of complete cycles or vibrations per unit of time —usually per second. Frequency of a wave is equal to the velocity divided by the wavelength.

*frequency drift:* An undesired change in the frequency of an oscillator or receiver.

*frequency stability:* The ability of an oscillator to maintain a predetermined frequency.

## G

*gain control:* A control that can change the overall gain of an amplifier. A volume control.

*ground wave:* A radio wave that is propagated near or at the surface of the earth.

## H

*half-wave antenna:* An antenna whose length is approximately equal to one half the wavelength to be received.

*ham:* A term applied to licensed amateur-radio operators.

*high frequency:* A frequency in the band extending from 3 to 30 MHz.

## I

*impedance:* The total opposition that a circuit offers to an alternating current or any other varying current at a particular frequency. A combination of resistance and reactance. The unit is the ohm.

*ionosphere:* The upper portion of the atmosphere of the earth beginning at about 30 miles above the surface.

## K

*kilohertz:* 1,000 hertz or cycles per second.

## L

*lead-in:* The conductor or conductors that connect the antenna to the receiver.

*lightning arrestor:* A protective device which leaks off static charges in the vicinity of an antenna to ground, thus it tends to prevent the charges from building up to the intensity of lightning.

*line:* Transmission line or power line.

*long waves:* Wavelengths longer than the longest broadcast-band wavelength of 545 meters. It is the frequency range from 60 to 540 kHz.

*low frequency:* A frequency in the band extending from 30 to 300 kHz in the radio spectrum.

# M

*medium frequency:* The band from 300 to 3000 kHz.

*megacycle:* One million cycles per second.

*maximum usable frequency (MUF):* The highest frequency reflected back to earth for a particular transmission path. The MUF fluctuates according to propagation conditions and the sunspot cycle.

*modulation:* The process in which the amplitude, frequency, or phase of a carrier wave is varied with time in accordance with the wave form of an intelligence signal.

# N

*noise filter (limiter):* A combination of one or more choke coils and capacitors to reduce man-made or other noises hindering reception.

# O

*ohm:* The practical unit of electrical resistance. It is that resistance across which 1 volt will cause a current of 1 ampere.

*omnidirectional:* In all directions, such as the radiation pattern of a vertical antenna.

# P

*piezoelectric:* Property of some crystals to generate a voltage when mechanical force is applied. Conversely, the ability to produce a mechanical force by expanding or contracting whenever a voltage is applied.

*power output:* The power in watts delivered by an amplifier to a speaker.

*propagation:* The travel of electro-magnetic waves or sound waves through a medium.

# Q

*QSL:* The acknowledgement requested by a listener from a station in answer to a reception report. Usually a full-color picture postcard with report data on the back.

# R

*radio spectrum:* The entire range of useful radio waves as classified into seven bands by the FCC.

| Designation | Abbr. | Frequency | Wavelength |
|---|---|---|---|
| very low frequency | vlf | 10-30 kHz | 30,000-10,000 m |
| low frequency | lf | 30-300 kHz | 10,000-1,000 m |
| medium frequency | mf | 300-3,000 kHz | 1,000-100 m |
| high frequency | hf | 3-30 MHz | 100-10 m |
| very high | | | |

| frequency | vhf | 30-300 MHz | 10-1 m |
| ultrahigh | | | |
| frequency | uhf | 300-3000 MHz | 100-10 cm |
| superhigh | | | |
| frequency | shf | 3000-30,000 MHz | 10-1 cm |

*reflected wave:* The sky radio wave, reflected back to earth from an ionosphere layer.

*refracted wave:* The wave that is bent as it travels into a second medium, as from the atmosphere into an ionized layer of the stratosphere.

## S

*S-meter:* A tuning indicator that provides an indication of the strength of the signal received. The meter is calibrated from 0 to 9. The receiver is tuned for the highest S-reading. The scale of some S-meters is also calibrated in decibels above S9.

*selectivity:* The ability of a receiver to reject undesired and untuned signals.

*sensitivity:* The ability of a receiver to pull in weak signals.

*shortwave:* From 1.6 MHz to 30 MHz.

*SINPO code:* A five-step quality report recognized by all international shortwave broadcasters.

*ssb* (single sideband) : A mode of transmission in which only one sideband, rather than both, is transmitted. A BFO (beat-frequency oscillator) and product detector are required for ssb reception.

## T

*telegraphy:* Communication by code signals sent over connecting wires.

# Abbreviations

a,A—approximate (frequency)
AA—Arabic
abt—about
ad—advertisement
addr—address
afc—automatic frequency control
AFRTS—American Forces Radio TV Service
agc—automatic gain control
am—amplitude modulation
ANARC—Association North American Radio Clubs
ancr—announcer
anl—automatic noise limiter
anmt—announcement
ant—antenna
approx—approximate
AP—A Portugeasa—Portuguese national anthem
ARO—amateur radio operator
avc—automatic volume control
BBC—British Broadcasting Corp.
BC, bc—broadcasting
bcb—broadcast band
BCC—Broadcasting Corp. of China
bcstr—broadcaster
bfo—beat-frequency oscillator
Br—British
BP—Boite Postale—post office box in French

CAm—Central America
CARACOL—Cadena Radial Colombiana
CB—Citizen's Band
CBC—Canadian Broadcasting Corp.
CE—chief engineer
cfm—confirm
ch—channel
ck—check
cl—close—signoff
c/l—call letters
cland—clandestine
CO—commanding officer
comcl—commercial
condx—conditions
Corp.—corporation
CP—Caixa Postal—post office box in Portuguese
cps—cycles per second
CQ—general call requesting contact
ctry—country
cw, CW—continuous wave—Morse code
cx—conditions
dB—decibel
dc—direct current
de—from
dem—democratic
df—direction finding

**DJ**—disc jockey
**dsb**—double sideband
**DU**—down under—Australia, New Zealand
**DX**—long-distance reception
**DXer**—one who listens for distant stations
**Ea**—east
**EE**—English language
**EN**—English language
**Eng**—engineer
**equip**—equipment
**es**—and
**EUR**—Europe
**exc**—except, excellent
**exp**—experimental
**fb**—fine
**FBIS**—Foreign Broadcast Information Service
**FCC**—Federal Communications Commission
**FE**—far eastern
**FF**—French, French speaking
**fm**—frequency modulation
**f/i**—fade-in
**f/o**—fade-out
**Fr**—French
**freq**—frequency
**f/up**—follow-up
**GC**—great circle
**GCP**—great circle path
**Ger**—German
**GMT**—Greenwich Mean Time
**GRS**—General Radio Service (Canada)
**gov**—government
**GSTQ**—God Save The Queen (British National Anthem)
**gud**—good
**ham**—amateur radio operator
**har**—harmonic
**het**—heterodyne
**hf**—high frequency
**HQ**—headquarters
**hrd**—heard
**id**—identification
**if**—intermediate frequency
**incl**—including, include
**info**—information

**ips**—inches per second
**IRC**—International Reply Coupon
**irreg**—irregular
**IS**—interval signal
**ITU**—International Telecommunication Union
**JJ**—Japanese, Japanese speaking
**kHz**—kilohertz
**kW**—kilowatt
**L**—language
**LA**—Latin America
**lf**—low frequency
**lgd**—logged
**lsb**—lower sideband
**LV**—la voz de . . . Spanish station identification
**lvl**—level
**lw**—longwave
**mb**—meter band
**ME**—Middle Eastern
**MHz**—megahertz
**mni**—many
**mw**—medium wave
**mwbc**—medium-wave broadcast
**mx**—music
**NA**—North America(n)
**NAf**—North Africa(n)
**Nac**—nacional—Spanish for national
**net**—network
**nil**—nothing
**nr**—near
**nx**—news
**Pac**—Pacific
**pgm**—program
**POD**—path of darkness
**pop**—popular music
**PP**—Portuguese—speaking
**prop**—propagation
**pse**—please
**ptp**—point-to-point (type of utility transmission)
**R**—radio
**rag**—magazine
**rcd**—received
**rcvr**—receiver
**re**—concerning
**Re**—report
**red**—rediffusion

rel—religious
rf—radio frequency
Rp—return postage
rpt—report
RTTY—radio teletype
Rx—receiver
S—unit of signal strength
SASE—self-addressed stamped envelope
sched—schedule(d)
scope—oscilloscope
sez—says
shud—should
sig—signal
SINPO—signal reporting code
sked—schedule
s/on—sign-on
s/off—sign-off
Sp—Spanish language
SS—Spanish speaking
stn—station
svc—service
sw—shortwave
swbc—shortwave broadcast

SWL—shortwave listener
TC—time check
t/i—tune-in
t/o—tune-out
tnx—thanks
UDX—utility DX
unid—unidentified
unk—unknown
urs—yours
v/c—verification card
verie—verification
v/l—verification letter
vert—vertical antenna
vfo—variable-frequency oscillator
v/s—person signing verification
vy—very
WCNA—west coast North America
wk—week, work, weak
wrp—weather report
wx—weather
xmsn—transmission
xmtr—transmitter
xtal—crystal
Z—Zulu time (GMT)

# Miscellaneous Codes

## Q-CODE

The following list of Q signals is not complete, but the signals are those most frequently sent. Most of the Q signals can be used in either question or statement form. You will hear these signals throughout the world as they are used frequently by ham operators.

**QRA**—What station are you?
**QRB**—How far are you from me?
**QRD**—Where are you headed and from where?
**QRE**—What is your estimated time of arrival?
**QRF**—Are you returning to _____?
**QRG**—What is my exact frequency?
**QRH**—Does my frequency vary?
**QRI**—How is the tone of my transmission?
**QRK**—How do you read my signals?
**QRL**—Are you busy?
**QRM**—Are you being interfered with?
**QRN**—Are you troubled by static?
**QRO**—Shall I increase power?
**QRP**—Shall I decrease power?
**QRQ**—Shall I speak faster?
**QRR**—Are you ready for automatic operation?
**QRS**—Shall I speak slower?
**QRT**—Shall I stop transmitting?
**QRU**—Have you anything for me?
**QRV**—Are you ready?
**QRW**—Shall I inform _____ that you are calling him on _____ kHz?
**QRX**—When will you call me again?
**QRY**—What is my turn?
**QRZ**—Who is calling me?

QSA—What is the strength of my signals?
QSB—Are my signals fading?
QSD—Are my signals mutilated?
QSG—Shall I send _____ messages at a time?
QSK—Can you hear me between your signals?
QSL—Will you send me a confirmation of our communication?
QSM—Shall I repeat the last message?
QSN—Did you hear me on _____ kHz?
QSO—Can you communicate with _____ direct or by relay?
QSP—Will you relay to _____ free of charge?
QSQ—Have you a doctor aboard?
QSR—Have the distress calls from _____ been cleared?
QSU—Shall I send reply on this frequency or on _____ kHz?
QSV—Shall I send a series of Vs on this frequency?
QSW—Will you send on this frequency?
QSX—Will you listen to _____ on kHz?
QSY—Shall I change to transmission on another frequency?
QSZ—Shall I send each word or group more than once?
QTA—Shall I cancel message number?
QTB—Do you agree with my counting words?
QTC—How many messages do you have for me?
QTE—What is my true bearing from you?
QTF—Will you give me the position of my station according to the bearings of your direction finding station?
QTG—Will you send two dashes of ten seconds each followed by our call sign—repeated _____ times on kHz?
QTH—What is your location in latitude and longitude?
QTI—What is your true track–in degrees?
QTJ—What is your true speed?
QTL—What is your true heading–in degrees?
QTM—Send signals to enable me to fix my bearing and distance?
QTN—At what time did you depart from _____?
QTO—Have you left port/dock?
QTO—Are you going to enter port/dock?
QTQ—Can you communicate with my station by means of the International Code of Signals?
QTR—What is the correct time?
QTS—Will you send your call sign for _____ minutes now, or at _____ hours on _____ kHz so that your frequency may be measured?
QTU—During what hours is your station open?
QTV—Shall I stand guard for you on _____ kHz?
QTX—Will you keep your station open for further communication with me for _____ hours?
QUA—Do you have news of _____?
QUB—Can you give me, in the following order, information concerning visibility, height of clouds, direction and velocity of ground wind at _____ (place of observation)?
QUC—What is the number (or other) of the last message you received from me?

**QUD**—Have you received the urgency signal sent by _____?
**QUF**—Have you received the distress signal sent by _____?
**QUG**—Will you be forced to alight (or land)?
**QUH**—Will you give me the present barometric pressure at sea level?
**QUJ**—Will you indicate the true course for me to follow?
**QUM**—Is the distress traffic ended?

## INTERNATIONAL MORSE CODE

Each code group shown is sent as a single symbol without pauses. The unit of code time is the *dit*, equivalent to a short tap on the telegraph key or a quick flip of the tongue. The *dah* is three times as long as a dit. Spacing between dits and dahs in the same character is equal to one dit. The spacing between letters in a word is equal to one dah. The spacing between words is equal to five to seven dits.

| *Letter/Number Punctuation* | *Code Symbols* | *Phonetic Sound* |
| --- | --- | --- |
| A | · — | ditdah |
| B | — · · · | dahditditdit |
| C | — · — · | dahditdahdit |
| D | — · · | dahditdit |
| E | · | dit |
| F | · · — · | ditditdahdit |
| G | — — · | dahdahdit |
| H | · · · · | ditditditdit |
| I | · · | ditdit |
| J | · — — — | ditdahdahdah |
| K | — · — | dahditdah |
| L | · — · · | ditdahditdit |
| M | — — | dahdah |
| N | — · | dahdit |
| O | — — — | dahdahdah |
| P | · — — · | ditdahdahdit |
| Q | — — · — | dahdahditdah |
| R | · — · | ditdahdit |
| S | · · · | ditditdit |
| T | — | dah |
| U | · · — | ditditdah |
| V | · · · — | ditditditdah |
| W | · — — | ditdahdah |
| X | — · · — | dahditditdah |
| Y | — · — — | dahditdahdah |
| Z | — — · · | dahdahditdit |
| 1 | · — — — — | ditdahdahdahdah |
| 2 | · · — — — | ditditdahdahdah |
| 3 | · · · — — | ditditditdahdah |
| 4 | · · · · — | ditditditditdah |
| 5 | · · · · | ditditditditdit |

| | | |
|---|---|---|
| 6 | — · · · · | dahditditditdit |
| 7 | — — · · · | dahdahditditdit |
| 8 | — — — · · | dahdahdahditdit |
| 9 | — — — — · | dahdahdahdahdit |
| Ø zero | — — — — — | dahdahdahdahdah |
| / fraction bar | — · · — · | dahditditdahdit |
| . period | · — · — · — | ditdahditdahditdah |
| ? question mark | · · — — · · | ditditdahdahditdit |
| , comma | — — · · — — | dahdahditditdahdah |
| error | · · · · · · · · | ditditditditditditditdit |
| : colon | — — — · · · | dahdahdahditditdit |
| ; semicolon | — · — · — · | dahditdahditdahdit |
| ( ) parenthesis | — · — — · — | dahditdahdahditdah |
| wait (AS) | · — · · · | ditdahditditdit |
| end of message (AR) | · — · — · | ditdahditdahdit |
| end of transmission (SK) | · · · — · — | ditditditdahditdah |